SEA-LEVEL CHANGES

FURTHER TITLES IN THIS SERIES

Elsevier Oceanography Series, 8

SEA-LEVEL CHANGES

by

EUGENIE LISITZIN

ELSEVIER SCIENTIFIC PUBLISHING COMPANY   *Amsterdam · Oxford · New York 1974*

ELSEVIER SCIENTIFIC PUBLISHING COMPANY
335 Jan van Galenstraat
P.O. Box 211, Amsterdam, The Netherlands

AMERICAN ELSEVIER PUBLISHING COMPANY, INC.
52 Vanderbilt Avenue
New York, New York 10017

1975

Library of Congress Card Number: 73-85225

ISBN 0-444-41157-7

With 50 illustrations and 67 tables

Printed in The Netherlands

ACKNOWLEDGEMENT

A work covering the different aspects of sea-level researches must necessarily be based on studies and results of many distinguished scientists who have worked or still work in this field. Some of these have already departed this life, which made it impossible for me to ask their permission to reproduce some of their figures or tables. On the other hand, it is a great pleasure for me to express my warm thanks to the following persons who kindly allowed me to use their results: Prof. Dr. A. Defant, Innsbruck; Prof. Dr. E. Palmén, Helsinki; Prof. Dr. W. Hansen, Hamburg and Mr. G.W. Lennon, Birkenhead.

Moreover, I should like to express my warmest gratitude for permission to reproduce the material from the publications of, at least, the following institutions and publishing companies: The Royal Society, London; Österreichische Akademie der Wissenschaften, Vienna; Direction du Service Hydrographique de la Marine, Paris; Deutsches Hydrographische Institut, Hamburg; American Geophysical Union, Washington, D.C.; Svenska Geofysiska Föreningen, Stockholm; Musée Océanographique de Monaco, Monaco-Ville; Council of the Institution of Civil Engineers, London; Springer Verlag, Heidelberg; and Pergamon Press Ltd, Oxford.

Helsinki, April 1973

Eugenie Lisitzin

CONTENTS

CHAPTER 1

INTRODUCTION

Oceanography is considered a young science with roots going back only to the first half of the nineteenth century. Sometimes as late a year as 1872, when the first scientific cruise of a modern nature, the famous "Challenger" Expedition, began its work in the oceans, is regarded as the opening year of oceanographic research. However, in this connection it must always be kept in mind that there is an important and interesting field within the boundaries of modern oceanography which has a considerably more respectable pedigree. This significant field consists of the studies on sea level and its variations. Research on the tides, especially on their theoretical aspects must, of course, be mentioned first. Nevertheless, there are other phenomena connected with sea-level changes which have been commonly known and studied for centuries. It may suffice to refer to two examples: the disastrous floods described, if not always in a scientific way, by many ancient peoples; and the land uplift characteristic of large areas in the northern hemisphere. The latter phenomenon has been known and studied, at least in the Fennoscandian countries, since the beginning of the eighteenth century. It gave, in the middle of the nineteenth century, the first impulse to the erection of sea-level measuring poles and thus laid the first firm foundation for purely scientific studies of sea-level changes, such as they appear in nature.

Sea-level research may at a first cursory glance be considered a rather unitary and well-limited field of scientific studies. The conclusion could easily be drawn that the contemporary tendency for specialization has created within the wide framework of oceanography a scientific branch which may allow the investigator to follow his own independent way. Nothing could be more erroneous than such an interpretation. It will be made clear, in the particular chapters of this book, that students of sea level and its variations are forced to consider in their work a considerable number of different elements, factors and phenomena which form a substantial part of many very different sciences. It may be sufficient to mention in this connection a few of these elements and phenomena. Hydrography of oceanography, in the more restricted sense of these terms, contribute such elements as temperature and salinity, and consequently also the density of sea water, currents and long waves; meteorology, atmospheric pressure, different wind effects, evaporation and precipitation; hydrology, water discharged from rivers; geology, land uplift and land subsidence; astronomy, gravitation and tide-generating forces; seismology, tsunami waves; and, finally, glaciology, the eustatic changes.

It may be of considerable interest to summarize as an introduction the different points of view presented by individual oceanographers on the classification of the causes for sea-level fluctuations. The principal purpose of this short survey is to emphasize the possibility of different approaches to the problem of the origin of sea-level variations (Lisitzin, 1972b).

One of the earliest summaries of the particular factors influencing sea-level dates from 1927 was presented by Nomitsu and Okamoto in a paper dealing with the causes of the seasonal fluctuations in sea level in the waters surrounding Japan. In their paper the two authors mentioned two principal groups of contributing factors. The first group refers to the internal causes, the second group to the external causes. The main characteristic of the internal causes is, according to Nomitsu and Okamoto, that they are connected with changes of the properties of the sea water. Besides the temperature and salinity of the sea water Nomitsu and Okamoto also ascribed to this group precipitation, evaporation and river discharge. To the group of external factors belong atmospheric pressure, the different effects brought about by the wind and the consequences of the Coriolis parameter upon the moving water masses. It may be of interest to point out that astronomical contribution to sea-level variation was not taken into consideration in the above classification.

Seventeen years after the first classification was presented, a paper on the changes in sea level in the Baltic Sea was published by Hela (1944). Hela also gave two principal groups characterizing the causes of sea-level variations and denoted them as the internal and the external causes. According to Hela, only the distribution in sea water temperature and salinity belongs to the former of these groups. Among the external factors Hela mentioned not only the tides but also the meteorologically conditioned elements, or, more precisely, atmospheric pressure, winds, seiches, precipitation, evaporation, river discharge and water transport through the transition regions. This classification seems to be adequate for many purposes and has been used in its original state or slightly modified in several different connections.

Nevertheless, efforts to create new classifications continued. Dietrich (1954) in a very interesting paper on sea-level variations at Esbjerg, Denmark, fitted the intrinsic elements into three large systems. The first of these systems covers the effect of the astronomical bodies upon the water in the oceans and seas; the second system concerns the ocean and the Earth's crust; and the third system deals with the ocean and the atmosphere. In this classification additional elements, such as the vertical movements of the Earth's crust and changes in the topography of the sea floor, are included in the second system. The third system covers, in addition to the meteorological factors, the hydrographic elements, since fluctuations in temperature and salinity of the sea water were considered by Dietrich to be the consequences of primarily meteorological effects.

A further attempt at classification of the causes of sea-level variations was made by Galerkin (1960). The author proposed, in his research on the seasonal cycle in sea level in the Sea of Japan, three principal sections of contributing factors. The first of these sections deals with the variations of the physical properties of sea water, which according to Galerkin are practically identical with the changes in water density. The second section covers the fluctuations in the quantity of water — which could therefore be characterized as 'water balance'. This section includes such factors as river discharge, precipitation, evaporation and water transport through the transition regions. The third section's contribution may at first appear to be fairly restricted, since it refers principally to the causes affecting the uneven distribution of sea-level heights within a basin. The constituents of this section are, however, very important factors in sea-level research, being atmospheric pressure, wind stress and Coriolis force.

As a general conclusion it may be pointed out that the development of the classifications has shown a more or less distinct transformation from a fairly ordinary to a more sophisticated division, thus reflecting the progress sea-level research has made during the decades concerned.

In spite of the particular advantages offered by the above classifications, it seemed preferable to select a quite different approach to the problem in the following description of the main features of the perpetual and continuous variations which are characteristic of the water surface in the oceans and seas. This procedure gives, in addition, a better opportunity to balance the extent of the separate chapters on the one hand, and on the other hand to pay more attention to problems which have mainly been discussed only in particular papers on specific questions and not in extensive compilation publications devoted either to different branches of oceanography or to the science as a whole. For instance, the theoretical background of the tides and the semi-diurnal and diurnal tidal constituents have been described fairly briefly in the following, since there are a large number of monographs on these subjects (cf., for instance, Sager, 1959; MacMillan, 1966; Godin, 1972). These questions are also thoroughly treated in many publications on general oceanography, e.g., Defant (1961, Vol.2, pp.244—516) and Dietrich (1963, pp. 394—474). However, relatively considerable space has in the following text been dedicated to the long-period tidal constituents, the description and characteristics of which are only found in compilation publications in exceptional cases.

There is a field which some readers may consider to be closely connected with sea-level research, but which has been almost completely left out of consideration in the book: that which refers to the instrumentation necessary for sea-level recordings. Of course, it cannot be denied that, in the earlier days of the rapidly developing researches into sea levels, devices for measuring the variations concerned were frequently designed by outstanding experts in this field. It may be sufficient to refer in this connection to Sir William Thomson, Witting, Renqvist, and Rauschelbach, although many more could be mentioned (Matthäus, 1972). The present development, aiming at a complete automatization of recording devices, has transferred the task of construction of sea-level recorders from scientifically trained oceanographers to technical specialists. The particular details connected with the design and construction of these devices are therefore hardly of any great interest to sea-level students. In addition, the proliferation of sea-level recorders developed during the last few years is so pronounced that a complete listing would require considerable space and would probably be incomplete. Moreover, many of the recently constructed devices have so far not proved their reliability for the intended purpose, at least not in the cases where high accuracy of the records is required.

The attentive reader will assuredly soon note that some parts of the water-covered areas and their coastal regions have been taken into account to a considerably higher degree than other regions. There are at least two different causes to which this regrettable fact may be ascribed. Firstly, it must always be kept in mind that the distribution of the sea-level recording gauges and tidal poles is extremely uneven. Thus there is, for instance, a fair amount of data available from most of the European coasts, from the United States of America and from Japan. Conversely, some other parts of the world oceans and their coastal regions are represented very poorly. There is no doubt that the lack of primary data must be reflected not only in the amount of reference literature, but also in the share allotted to the particular regions in this text. Secondly, the author must confess that since her home country, Finland, is bordered by the blue waves of the Baltic Sea, her main interest and − why not declare it − her principal duty during a prolonged span of years has been dedicated to the study of sea-level variations and associated phenomena characteristic of this sea basin. The author is self-evidently aware of the fact that extensive parts of the world oceans have been unfairly treated in the following chapters. However, it must always be remembered that, since all oceans and seas are interconnected, sea-level changes in one part of the Earth's globe must respond to related fluctuations in other, possibly relatively distantly situated regions. In addition, the methods of computation used for one sea basin may frequently, although perhaps with some slight modifications, be utilized for other aquatic areas. The author has in many cases had the advantage and pleasure of benefits from the research work done by other scientists who are specialists in the field of sea-level studies, and would like to express in this connection her warmest thanks to these distinguished oceanographers.

Finally, it must be pointed out that the Baltic Sea is a highly interesting research region as regards sea-level variations. For instance, since the tidal phenomenon is rather insignificant in this sea area, the effect of other contributing factors upon the sea level may be studied without the disturbances due to astronomically caused variations. In addition, the Baltic Sea may, at least in some respects, be considered as a natural laboratory or a model basin of large proportions. All these facts have been recognized by Finnish scientists and also by the Finnish government for a long span of years. There has been a special department for sea-level research at the Institute of Marine Research in Finland for more than half a century. Reference may also be made in this connection to Rolf Witting, the first director of this institute, who during the first quarter of this century was not only a name but also a personality well-known to most oceanographers of those days. His interest in sea-level research was pronounced and was by no means restricted to the Baltic. Witting was the first person to propose the establishment of the International Committee on Mean Sea Level, which during a long span of years has performed much valuable work. During the 1920's and 1930's the names of the Finnish oceanographers concerned with different aspects of sea-level research, e.g., Henrik Renqvist, E. Palmén and S.E. Stenij, belonged to the most outstanding of the day even in international circles. Unfortunately, times and aspirations are subject to changes. Today the position of sea-level studies in Finland is not as favourable as it was during the years before the spring of 1972. The author of this book has had her most active period before this critical time and has therefore no excuse. It is up to the reader to express his or her opinion of the efforts made and the results achieved as described in the following pages.

CHAPTER 2

PERIODICAL SEA-LEVEL CHANGES

ASTRONOMICAL TIDES

Tidal theory – semi-diurnal and diurnal tides

The study of the phenomena connected with astronomical tides is the oldest purely scientific branch, not only of sea-level researches, but also of all oceanographic investigations. The roots of scientific tidal studies go back as far as the seventeenth century. The foremost place of honour belongs in this respect to Sir Isaac Newton, who in his famous work *Philosophiae Naturalis Principia Mathematica*, published in 1687, laid the first firm foundation for a mathematical investigation of the tides. Additional mathematical and physical explanations of these phenomena were given during the first part of the eighteenth century by Bernoulli, Euler and MacLaurin. Some hundred years after Newton's epoch-making work appeared, the study was continued by Laplace, while the names connected with tidal research during the nineteenth century were Lord Kelvin (Thomson) and Poincaré. These distinguished scientists also laid the first basis for the treatment of the tidal phenomena as a practical problem.

Newton's great achievement was the discovery of the laws of gravitation. This discovery allowed the explanation of the tidal phenomena as the consequence of the attraction exerted by the Sun and Moon upon the water particles in the oceans and seas. Newton also developed the equilibrium theory of the tides while being, however, conscious of the fact that this theory was only a rough approximation of the phenomenon concerned. Starting from this foundation, Laplace, Kelvin and others developed the dynamic theory of tides.

Equilibrium tides are understood as the tides which would occur in a non-inertial ocean covering the whole Earth-globe. Many features related to the oceanic tides may be explained by the equilibrium theory, but a comparison with the observations indicates that there are also a number of considerable deviations. Although spring tides appear around the time of full moon and new moon, and neap tides at the quadratures, and the heights of spring tides are considerably higher that those of neap tides, the observed tides show amplitudes which are generally much greater than those derived from the equilibrium theory. According to the dynamic theory developed by Laplace, the problem of the tides is one of motion, not a static problem. The dynamic theory stipulates that tides are waves caused by rhythmical forces and they are therefore characterized by the same periods as these forces. For the final development of tidal waves, such as they appear in nature, factors other than the tide-generating forces must be taken into consideration. Among these factors reference may, for instance, be made to the depth and the configuration of the ocean or sea basin, the deflecting force of the Earth's rotation (Coriolis force) and frictional effects of differing kinds. Since the tide-generating forces are known with great accuracy, the hydro-dynamic equations representing the motion of the water particles may be derived. The first equations of this type were presented by Laplace. However, the general equations of the dynamic theory have not been solved yet, in so far as the tides are concerned.

For the practical examinations of tidal phenomena the harmonic theory of tides has been developed. The starting hypothesis of this theory of tides is similar to that for the dynamic theory: that the tidal fluctuations must be characterized by the same periods as the tide-generating forces. Through the harmonic theory the basis was presented not only for the understanding of numerous tidal phenomena, but also for their prediction in time and space.

In the latter part of the nineteenth century and during the first part of the twentieth century tidal research made considerable progress and contributed markedly to the knowledge of tidal phenomena. Among the leading scientists in this field in Great Britain must be mentioned, in addition to Lord Kelvin, Sir George Darwin, J. Proudman, A.T. Doodson and their foremost successor, the late J.R. Rossiter (†1972). In Germany and Austria the leading names were A.Defant, R. von Sterneck and H. Thorade, and in the United States R.A. Harris and H.A. Marmer. To a younger, still active generation of specialists on different aspects of tidal research, belong W.H. Munk and B.D. Zetler in the United States. The number of publications dedicated every year to different tidal problems is accelerating. It may therefore be appropriate to refer to the most comprehensive existing bibliographies on tides. They have been published by the International Association of Physical Oceanography (Association d'Océanographie Physique, 1955, 1957a, 1971b). The three volumes on tidal bibliography cover a time-period of over 300 years, extending from 1665 to 1969.

Before proceeding to a more detailed study of the harmonic theory of tides and its practical applications, a few words must be devoted to the general significance of the equilibrium theory. Doodson (1921) has compiled the amplitudes and angular speeds of all the tidal constituents which may be determined on the basis of the gravitational theory of tides. The harmonic units of the equilibrium tides are known with great accuracy and in some cases they have been utilized in tidal research. Nevertheless, it must always be kept in mind that the equilibrium theory may be applied only as a first approximation and exclusively in deep, open oceanic regions, while in shallow water and in the vicinity of the coasts the behaviour of the particular tidal constituents deviates to a very pronounced degree from the somewhat simplified features which are characteristic of the equilibrium tides.

For practical studies connected with the character of the tides in different oceans and seas, as well as for tidal prediction with navigational purposes in mind, a completely different approach to the problem is necessary. The method used in this connection consists of utilizing the tidal observations or records made at a given locality for the forecast of the tide for any selected period in the future. This manner of procedure has yielded valuable results. The greatest disadvantage of this method is that, self-evidently, it can be utilized only for such localities for which previous tidal data are already available. Only the frequencies of the particular harmonic tidal constituents are determined on the basis of the knowledge of the tide-generating forces. The amplitudes and the phase angles for all tidal constituents must be determined from the observed data. The final result, representing the general features of the tidal phenomenon characteristic of the locality concerned, is reached by summing up a sufficient number of the harmonic tidal constituents.

According to Newton's law of gravitation, the gravitational attraction between two astronomical bodies is directly proportional to the product of their masses and inversely proportional to the square of the distance between them. The formula for the gravitational force F may thus be expressed in the following way:

$$F = \frac{\gamma\, m_1 m_2}{r^2}$$

where m_1 and m_2 are the masses of the two bodies separated by the average distance r and γ is the so-called constant of gravitation. In order to determine the gravitational force existing between two bodies the particular components of the force must be integrated over the total of the mass elements of these bodies. For bodies where the distribution of mass is not uniform, the equation given above will, of course, result in approximate values only. The values of the gravitational force will be the more accurate the greater is the distance between the bodies compared with their dimensions.

If the Moon and the Sun attracted every water particle in the oceans and seas with the same force, there would not be any tides. It is the extremely small but perceptible deviation in the direction and magnitude of the gravitational force of the two celestial bodies upon the particular points on the Earth's surface which is the cause of the tidal stresses and the tidal phenomena, such as they are observed in nature.

Fig. 1 illustrates schematically the effect of the lunar gravitational force upon different

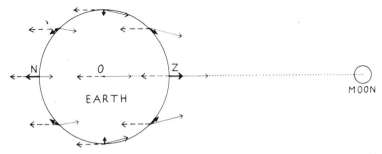

Fig. 1. The tide-generating force resulting from the attractive and the centrifugal forces. At Z the Moon is in zenith, at N it is at nadir. Light solid arrows represent the attractive force of Moon, dashed arrows the centrifugal force and heavy solid arrows the tide-generating force.

points on the Earth. At the point Z the Moon is in the zenith and at the point N it is at the nadir. Owing to the difference in distance the upward-directed force of the lunar attraction is somewhat greater at point Z than the downward-directed force at point N. In a corresponding way attractive forces deviating in magnitude cause stresses on every part of the Earth's surface. The gravitational attraction of the Moon upon the Earth corresponds to the vector sum of a constant force represented by the lunar attraction on the Earth's centre and a small deviation which for every point on and in the Earth depends on the distance from the Moon. It is this slight deviation which is called the tide-generating force. The larger constant gravitational force is counterbalanced by the centrifugal force of the Earth in its orbital rotation around the centre of the mass system represented by Earth and Moon, and it may therefore be left out of consideration in connection with the investigations of all tidal phenomena. Conversely, the tide-generating forces form the basis for the knowledge of the character and distribution of the tidal constituents over the surface of the Earth.

The tide-generating force may easily be computed for zenith, the centre of the Earth and nadir. If a is the radius of the Earth and r the distance between the centre of the Moon and that of the Earth, m the mass of the Moon and μ an element of the mass of the Earth at the point under consideration, we arrive at the following values for the different points:

	Zenith	Centre of the Earth	Nadir
The force of attraction	$\gamma \dfrac{\mu m}{(r-a)^2}$	$\gamma \dfrac{\mu m}{r^2}$	$\gamma \dfrac{\mu m}{(r+a)^2}$
The centrifugal force, corresponding at the Earth's centre to the negative force of attraction	$-\gamma \dfrac{\mu m}{r^2}$	$-\gamma \dfrac{\mu m}{r^2}$	$-\gamma \dfrac{\mu m}{r^2}$
The tide-generating force	$\gamma \mu m \left[\dfrac{1}{(r-a)^2} - \dfrac{1}{r^2}\right]$	0	$-\gamma \mu m \left[\dfrac{1}{r^2} - \dfrac{1}{(r+a)^2}\right]$
or neglecting higher terms containing a	$\gamma \mu m \dfrac{2a}{r^3}$	0	$-\gamma \mu m \dfrac{2a}{r^3}$

The above-given values of the tide-generating forces are the maxima which can be found on the Earth's surface. They show that the tide-generating force is proportional to the mass of the perturbating body and inversely proportional to the cube of the distance of the Earth to this body. In the hemisphere facing the Moon or Sun the force is directed towards the perturbating bodies, in the opposite hemisphere it acts away from them. The significance of the inverse cube in comparison with the inverse square in the equation for the gravitational force is distinctly shown by the fact that the effect of the Moon, in so far as the tidal phenomenon is concerned, is 2.17 times larger than that of the Sun, while

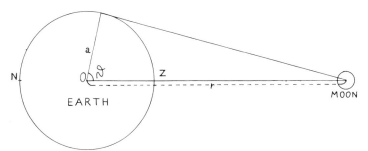

Fig. 2. The basis for the determination of the tide-generating potential.

the direct solar gravitational attraction on the Earth's surface is approximately 180 times larger than the lunar attraction.

The tide-generating force for every point on the Earth may be expressed as the gradient of the tide-generating potential W and as a function of the zenith distance ϑ of the Moon in the following way (Fig. 2):

$$W = \frac{1}{2} \gamma \frac{ma^2}{r^3} (3 \cos^2 \vartheta - 1) \tag{1}$$

where the symbols have the same significations as above. It may be pointed out in this connection that W is symmetrical in respect to the Earth–Moon axis, depending on the variable ϑ.

In a non-inertial ocean covering the entire surface of the Earth, the elevation $\overline{\xi}$ of the equilibrium tide is determined as a function of the Earth's own gravity and the tide-generating forces. In this case we have the equation:

$$\overline{\xi} = \frac{W}{g} + \text{constant} \tag{2}$$

where W is determined for the surface of the Earth and g stands for the acceleration of the Earth's gravity. The constant term in the equation ensures that the volume of the masses involved in the process remains unchanged. Only in the case of a global ocean is the constant zero.

For the harmonic analysis of the tidal variations of different types it is convenient to express the equilibrium tide as the sum of three terms:

$$\overline{\xi} = \frac{3}{4} \frac{\gamma m}{g} \frac{a^2}{r^3} [(3 \sin^2\delta - 1)(\cos^2 \theta - 1/3) + \sin 2\delta \sin 2\vartheta \cos(\alpha + \phi)$$

$$+ \cos^2 \delta \sin^2 \theta \cos 2(\alpha + \phi)] \tag{3}$$

In this equation the signification of the terms γ, m, a, g and r is given in connection with the eq. 1 and 2, while θ is the co-latitude, and ϕ the longitude east, δ the declination and α the west hour angle of the Moon, counted from Greenwich.

Eq. 3 shows the essential properties of the tidal elevation varying with time, but it is not entirely satisfactory, since both the declination and the distance between the Earth and the Moon are variable with time. A complete harmonic analysis of the tidal elevation requires eq. 3 to be expanded in a series of cosine and sine functions, with constant amplitudes and constant periods. However, for a general survey of the character of the tidal phenomenon, eq. 3 is sufficient.

The first term in this expression represents a tidal constituent which is independent of the longitude. The so-called long-period tides, to be described in the following section (pp. 37–51) arise from this term.

The second term of eq. 3 is a tidal constituent which at any instant has a maximum elevation at the latitudes 45°N and 45°S on the opposite sides of the equator. As a consequence of the factor $\cos(\alpha + \phi)$ the tides move in a westerly direction in relation to the Earth. During this rotation every geographical point performs a complete cycle during a lunar day. Owing to the factor $\sin 2\delta$ the diurnal tide is, according to the equilibrium theory, zero when the Moon crosses the equator. Because of the factor $\sin 2\theta$ there is no diurnal equilibrium tide at the equator and at the poles.

Considering the third term of eq. 3, it may be noted that it represents a tidal constituent which at any instant has two maximum elevations on the equator situated at the opposite sides of the Earth. These maxima on the equator are separated by two minima elevations. The whole system is moving westward relative to the Earth and a complete cycle is also in this case completed during a lunar day. The difference in respect to the diurnal constituent, represented by the second term, is that owing to the $\cos 2(\alpha + \phi)$ factor every geographical point on the Earth's surface is characterized by two complete cycles during this time. The constituents of this type of tide are therefore called the semi-diurnal tides. The effect of the factor $\sin^2 \theta$ is that no semi-diurnal equilibrium tide occurs at the poles, while the tidal range reaches the most pronounced values at the equator.

There are several cases where it is possible to determine the elevations for particular oceans, although the constant term in eq. 2 is not zero. The designation 'corrected equilibrium tide' is introduced in such cases. The uncorrected and the corrected equilibrium tides have been of considerable significance for the development of the harmonic theory of tides. However it must be pointed out again, that these tides, based on the assumption of a non-inertial motion, may be taken into consideration in nature only as an approximation and for tidal constituents with a period exceeding one year.

The solar tides may be determined following the same principles. Also in this case there are three different types of tidal constituents: long-period, diurnal and semi-diurnal. The equilibrium tide is the sum of both the lunar and the solar tides. At new moon and at full moon, when Sun and Moon are approximately in the same position, the range of the tide is at its highest, since the two systems of tides reinforce each other. At the quadratures the solar effect counteracts to some extent the lunar effect since the principal constituents of the two systems are out of phase.

Eq. 3 extended to cover all tidal constituents offers the possibility of determining the tidal potential and elevation as the sum of sine- and cosine-terms with a constant amplitude and frequency. The position of the Moon and Sun with respect to the Earth is a function of the distance from the Earth's centre and the latitude and longitude measured with respect to the ecliptic. These three factors are periodic functions of the following five angles:

s = the mean longitude of the Moon,
h = the mean longitude of the Sun,
p = the longitude of the perigee of the Moon's orbit,
N = the mean longitude of the ascending node of the Moon's orbit, $N = -N'$,
p_s = the longitude of the perigee of the Sun's orbit.

In Table I are collected the values of the changes σ° of these five angles during a mean solar hour and the periods in solar days or years.

Doodson (1921) developed the potential into single harmonic constituents of the equilibrium tide. The principal characteristic of this system is that it gives not only the angular speeds of hundreds of tidal constituents, but also their amplitudes. A considerable number of these constituents are of no practical significance and they may therefore be left out of consideration here. Some of the more important tidal constituents are collected in Table II. In this table the first number in the column designated 'Number' indicates the approximate number of tidal cycles per day. The remaining numbers represent a special notation of the arguments according to a scheme elaborated by Doodson. The number as a whole thus serves to denote the argument and may also be used to denote the constituent.

The long-period constituents Sa and Ssa represent the solar annual and semi-annual tides, respectively. Mm and Mf are the lunar monthly and fortnightly tides. All the diurnal constituents depend on the variation of the declination of Moon and Sun. K_1 is the most pronounced of all the diurnal tides and it is associated with the variation of both declinations. O_1 is lunar in origin, while P_1 is solar. Q_1 and J_1 are due to the changing distance of the Moon from the Earth. Among the semi-diurnal tides the principal lunar tide M_2 is the most dominant constituent, next followed by the principal solar tide S_2. N_2 and L_2 are tidal constituents due to the ellipticity of the Moon's orbit. T_2 is the corresponding

TABLE I

VALUES CHARACTERIZING FIVE ASTRONOMICAL ANGLES

Angle	σ°/hour	$360^\circ/\sigma^\circ$	
s	0.549017	27.321582	days
h	0.041069	365.242199	days
p	0.004642	8.847	years
N	−0.002206	18.613	years
p_s	0.000002	20,940	years

TABLE II

SOME TIDAL CONSTITUENTS AND THEIR CHARACTERISTICS

Number	Symbol	Argument	Speed degrees/hour	Relative coefficient
Long-period tides				
055.565	M_N	N'	0.0022	0.0655
056.554	Sa	$h - p_s$	0.0411	0.0118
057.555	Ssa	$2h$	0.0821	0.0729
063.655	MSm	$s - 2h + p$	0.4715	0.0158
065.455	Mm	$s - p$	0.5444	0.0825
073.555	MSf	$2s - 2h$	1.0159	0.0137
075.555	Mf	$2s$	1.0980	0.1564
075.565	Mf_N	$2s + N'$	1.1002	0.0648
085.455	Mt	$3s - p$	1.6424	0.0300
085.465	Mtt	$3s - p + N'$	1.6446	0.0124
Diurnal tides				
135.655	Q_1	$15°t + h - 3s + p - 90°$	13.3987	0.0722
145.555	O_1	$15°t + h - 2s - 90°$	13.9430	0.3769
163.555	P_1	$15°t - h - 90°$	14.9589	0.1785
165.555	K_1	$15°t + h + 90°$	15.0411	0.5305
175.455	J_1	$15°t + h + s - p + 90°$	15.5854	0.0296
185.555	OO_1	$15°t + h + 2s + 90°$	16.1391	0.0162
Semi-diurnal tides				
235.755	$2N_2$	$30°t + 2h - 4s + 2p$	27.8953	0.0230
237.555	μ_2	$30°t + 4h - 4s$	27.9682	0.0278
245.655	N_2	$30°t + 2h - 3s + p$	28.4397	0.1739
247.455	ν_2	$30°t + 4h - 3s - p$	28.5126	0.0330
255.555	M_2	$30°t + 2h - 2s$	28.9841	0.9081
265.455	L_2	$30°t + 2h - s - p + 180°$	29.5285	0.0257
272.556	T_2	$30°t - h + p_s$	29.9589	0.0248
273.555	S_2	$30°t$	30.0000	0.4236
275.555	K_2	$30°t + 2h$	30.0821	0.1151

solar tide. K_2 is the equivalent to K_1 in the group of diurnal tides and is thus associated with the variation of declination of both Moon and Sun.

It has already been mentioned above that the equilibrium theory of the tides cannot as such be utilized for the determination and prediction of the tides at a given locality. In these cases we have always to depend upon the observed or recorded sea-level data. The constituents of the actual tide differ in phase with respect to those of the equilibrium tide by a lag, which must be determined for each constituent and for every station on the basis of observations. Also, the amplitudes of each tidal constituent have to be computed with the help of observed data.

In Table I there was listed N, the mean longitude of the ascending node of the lunar orbit, with the period covering 18.61 years. This variation affects the declination and other factors. This variation must always therefore be included in the harmonic constituents by adding the nodal factor f and the nodal angle u corresponding to the nodal period. In this way is obtained for every tidal constituent an expression of the form:

$$f H \cos (V_0 + \sigma t + u - \kappa) \tag{4}$$

where σ is the angular speed, expressed in degrees per solar hour, V_0 corresponds to the starting instant of the computations, t is the time, usually given in the standard time zone of the particular locality of observation. H and κ are respectively the amplitude and the phase which, as has already been pointed out above, must be determined separately for each locality by means of direct observations. They are called the harmonic constants.

The introduction of f and u in eq. 4 indicates that the analysis of the more important tidal constituents in the oceans should always cover a period corresponding to the revolution of the node of the lunar orbit, i.e., approximately 19 years. In practice, principally as a consequence of the considerable work involved in the analysis of the tidal data and the high standard required in the tidal observations themselves, an analysis covering such a prolonged time is generally not feasible. Usually, a period of one year is sufficient to provide practically acceptable results. Different schemes have been developed for the practical execution of the harmonic analysis of tidal data based on periods of different length.

Table II shows that the relative coefficients for the two semi-diurnal tidal constituents M_2 and S_2 are the most pronounced not only in the particular group, but also of all the constituents. These two tides are responsible for the most commonly occurring type of tidal fluctuations in the oceans – the semi-diurnal tide with two high waters and two low waters per day. The speed difference between the two constituents results in their periods deviating by 25 minutes, the periods themselves being 12 h 25 min and 12 h respectively. This difference brings about the main features of the semi-diurnal tide in the oceans: spring tides and neap tides. Spring tides are called the tides within a semi-lunar period of 15 days which have the greatest range, i.e., the greatest difference between high water and low water. They should occur for the days of new moon and full moon, when the gravitational effects of Moon and Sun reinforce each other, but in practice this is by no means the case. Neap tides are the tides which occur near the time of the first and third quadratures of the Moon, they are characterized by the least marked range, since Moon and Sun, being in opposition, have counteracting effects. In addition, the contributing effects of all the other semi-diurnal constituents cause deviations not only in the range but also in the period of the semi-diurnal variations during a tidal spring–neap cycle. In the cases where the period between two high waters is greater than the lunar period of 12 h 25 min the term lagging tide is used. If the period is less than the lunar period the corresponding term is priming tide.

There are also considerable seasonal variations in the range of the semi-diurnal tides, especially pronounced in localities where the sea-level variation is large during the day. The greatest ranges, usually associated with the occurrence of the highest and lowest sea levels, are generally observed near the time of the solstices, i.e., in June and December. During spring and autumn, close to the time of the equinoxes, the semi-diurnal inequality is, as a rule, less pronounced.

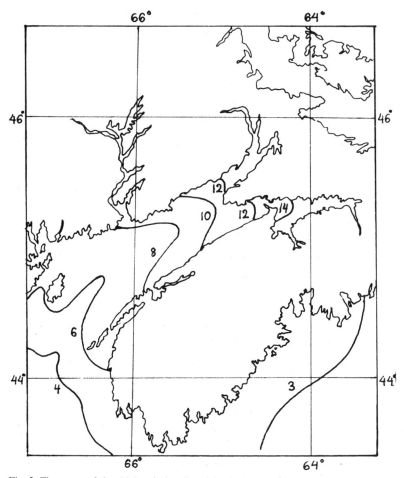

Fig. 3. The range of the tidal variation (in m) in the Bay of Fundy (Voit, 1956).

The most marked tidal range so far observed has been noted in the Bay of Fundy on the Atlantic coast of North America (Fig. 3), where the tidal variation exhibits ranges exceeding 15 m. Other fairly pronounced ranges have been observed in the Gulf of St. Malo, having sea-level differences of more than 12 m, and in the Bristol Channel, where the range exceeds 11 m. All these considerable ranges are considered to be caused by the resonance of the semi-diurnal constituents with the oscillation of the basins themselves. The continuous narrowing of the cross-section in the bays is assuredly in some cases an additional factor for the increase in range. In the oceans the tidal ranges never reach such marked proportions. In some localities in the South Pacific Ocean, the Arctic Ocean and the Mediterranean Sea the tidal range does not exceed 50–60 cm. In this connection it may also be mentioned that the diurnal tide is not much more pronounced in the Bay of Fundy and in the Gulf of St. Malo than in the oceans.

Photograph 1. High water at St. Malo at 08h43, September 6, 1963. The sea-level height is 12.50 m. The picture is taken towards the northwest. (Photograph: Service Hydrographique de la Marine. Paris.)

Photographs 1 and 2 show the difference in sea level between high and low water at St. Malo. The photographs were taken on September 6, 1963, by the Service Hydrographique de la Marine. The former of these photographs refers to the time 08h43 and a sea-level height of 12.50 m, the latter to the time 15h46 and the sea level of 0.85 m. The sea-level

Photograph 2. Low water at St. Malo at 15h46, September 6, 1963. The sea-level height is 0.85 m. The picture is taken towards the northwest. (Photograph: Service Hydrographique de la Marine, Paris.)

difference is thus almost 12 m. The pictures are taken towards the northwest. The two rocks seen on the photos are the Grand Bé to the right and the Petit Bé to the left. In the distance between the rocks is seen the island of Cézembre.

The range of the tidal constituents deviates considerably in different parts of the oceans. Along the coasts there have also been observed differences which may be rather small-scale in character. These differences may, however, be the consequence of the selection of the localities for the erection of the tide-measuring gauges, which in some cases are erected along the open coast and in other cases are situated on estuaries and rivers. It is a well-established fact that the range of the tide changes considerably as soon as the tidal wave moves up-river.

There are also some other peculiarities which have been noted in connection with the semi-diurnal tides. In general, and in agreement with the theoretical requirements, the range of the lunar semi-diurnal constituent M_2 is more than twice as large as that of the semi-diurnal constituent S_2. Nevertheless, along the coast of southern Australia the response to the solar tide is at some localities more marked than that to the lunar constituent. As a result high water may be observed there during several successive days at the same hour, instead of the generally more common daily retardation of approximately 50 min.

Table III gives the harmonic constants of the two principal semi-diurnal constituents M_2 and S_2 and the two main diurnal constituents K_1 and O_1. Most of the data are taken from the extensive work by Defant (1961, Vol. 2, pp. 364–503). Besides the tidal data, of which those reproduced in Table III are only a selection, Defant gives a considerable amount of additional information about the tidal phenomenon in different oceans and seas. This description is, moreover, in numerous cases illustrated by charts.

As already mentioned above, the data in Table III are only a small part of all available tidal data. The selection was difficult, since the quantity of data had to be restricted. However, special attention was paid to different types of tides, for instance, to the pronounced deviations between the amplitudes of the particular constituents depending on the location of the tidal stations. In order to give an example it may be mentioned that the tides are considerably weaker in the middle parts of the Pacific Ocean, represented by the five island groups whose harmonic constants are reproduced at the end of Table III, than in the coastal regions of this ocean. This feature has already been referred to above. The extremely marked differences between the range of the bays (St. Malo and Cardiff), on the one hand, and the more-or-less enclosed sea basins such as the Baltic Sea (Karlskrona, Libau, Helsinki, Ratan) and the Mediterranean (Genoa, Palermo, Trieste, Port Said), on the other hand, may also be emphasized.

The explanation of the character of the tides in bays and near-landbound seas of more limited dimensions needs in numerous cases the introduction of such terms as friction and Coriolis parameter in order to reach satisfactory results. Since the effect of friction increases with increasing amplitudes, the period of the free oscillation increases too. Friction may thus counterbalance the occurrence of a total resonance. The influence of the Coriolis parameter may cause oscillations which are perpendicular to the direction of

TABLE III

THE HARMONIC CONSTANTS OF THE MAIN TIDAL CONSTITUENTS FOR A NUMBER OF LOCALITIES

Locality and position		M_2	S_2	K_1	O_1	$\dfrac{K_1 + O_1}{M_2 + S_2}$	$M_2 + S_2 + K_1 + O_1$
Bergen	H(cm)	43.9	15.9	3.2	3.2	0.11	66.2
60°24′N 06°18′E	$\kappa°$	295	334	171	18		
Hornbaek		6.8	2.3	0.8	2.9	0.41	12.8
56°06′N 12°28′E		260	224	347	337		
Copenhagen		6.0	2.7	0.5	2.0	0.29	11.2
55°41′N 12°36′E		279	250	37	7		
Gedser		3.6	0.8	1.6	1.8	0.77	7.8
54°34′N 11°58′E		192	207	180	143		
Karlskrona		0.6	0.3	1.1	1.4	2.85	3.4
56°06′N 15°35′E		119	136	137	213		
Libau		0.1	0.2	0.6	0.6	3.71	1.5
56°32′N 20°59′E		128	110	297	277		
Helsinki		0.4	0.2	1.7	1.8	6.38	4.1
60°09′N 24°58′E		186	240	33	16		
Ratan		0.2	0.1	1.4	1.1	7.54	2.8
64°00′N 20°55′E		15	346	7	321		
Esbjerg		65.7	16.0	4.9	8.6	0.17	95.2
55°29′N 08°27′E		59	122	81	288		
Hook of Holland		75.3	18.6	7.9	11.2	0.20	113.0
51°59′N 04°07′E		71	131	351	181		
Dunkirk		206.8	62.2	3.9	6,5	0.04	279.4
51°03′N 02°22′E		358	50	355	170		
St. Malo		374.6	148.9	9.3	8.0	0.03	540.8
48°38′N 02°02′W		176	225	91	342		
Southend		200.7	57.5	11.0	12.0	0.09	281.6
51°31′N 00°43′E		355	50	6	193		
Lowestoft		71.0	21.3	17.7	14.0	0.34	124.0
52°29′N 01°46′E		264	324	334	167		
Aberdeen		130.9	45.0	11.0	12.7	0.13	199.8
57°09′N 02°05′W		21	58	203	50		
Sound of Sura		46.0	25.0	8.5	8.2	0.24	87.7
56°03′N 05°38′W		98	132	183	33		
Greenock		132.8	31.6	5.0	7.3	0.07	176.7
55°57′N 04°46′W		337	42	224	54		
Liverpool		305.5	97.1	11.9	11.1	0.06	425.6
53°25′N 03°00′W		320	5	190	42		
Cardiff		409.0	142.0	9.4	6.7	0.03	567.1
51°27′N 03°10′W		191	239	144	6		
Brest		203.5	74.3	6.4	6.7	0.05	290.9
48°23′N 04°30′W		100	139	72	323		

TABLE III (continued)

Locality and position		M_2	S_2	K_1	O_1	$\dfrac{K_1 + O_1}{M_2 + S_2}$	$M_2 + S_2 +$ $K_1 + O_1$
Lisbon	H(cm)	118.3	40.9	7.4	6.5	0.09	173.1
38°42′N 09°08′W	$\kappa°$	60	88	51	310		
Cadiz		92.8	32.3	6.1	6.1	0.10	137.3
36°30′N 06°12′W		41	69	28	294		
Genoa		9.0	3.1	3.2	1.2	0.36	16.5
44°25′N 08°55′E		240	256	188	190		
Palermo		10.9	4.5	3.1	1.1	0.27	19.6
38°08′N 13°20′E		264	285	213	135		
Trieste		26.3	15.8	17.5	5.0	0.53	66.6
45°39′N 13°46′E		277	285	70	57		
Port Said		11.7	6.9	2.1	1.7	0.20	22.4
30°15′N 32°18′E		304	319	305	275		
Ponta Delgada (Azores)		49.1	17.9	4.4	2.5	0.10	73.9
37°44′N 25°40′W		12	32	41	292		
Funchal (Madeira)		71.2	26.7	6.1	4.5	0.11	108.5
32°38′N 16°55′W		10	31	29	285		
Las Palmas		76.0	28.0	7.0	5.0	0.12	116.0
28°09′N 15°25′W		356	19	21	264		
Freetown		97.7	32.5	9.8	2.5	0.09	142.5
08°30′N 43°14′W		201	234	334	249		
Takoradi		45.9	15.4	11.8	2.0	0.23	75.1
04°54′N 01°45′W		100	127	347	321		
Cape Town		48.6	20.5	5.4	1.6	0.10	76.1
33°54′S 15°25′E		45	88	127	243		
Port Louis (Falkland)		47.1	15.0	10.9	13.7	0.40	86.7
51°33′S 58°09′W		157	195	37	4		
Buenos Aires		30.5	5.2	9.6	15.4	0.70	60.7
34°36′S 58°22′W		168	248	14	202		
Rio de Janeiro		32.6	17.2	6.4	11.1	0.35	67.3
22°54′S 43°10′W		87	97	148	87		
Port of Spain (Trinidad)		25.2	8.0	8.8	6.7	0.47	48.7
10°39′N 61°31′W		119	139	187	178		
Nassau (Bahama Is.)		37.9	6.4	8.7	6.5	0.34	59.5
25°05′N 77°21′W		213	237	120	124		
St. Georges (Bermudas)		35.5	8.2	6.4	5.2	0.27	55.3
32°22′N 64°42′W		231	257	124	128		
Fernandina		87.0	15.5	10.5	7.7	0.18	120.7
30°41′N 81°28′W		228	258	127	129		
Halifax		63.1	14.4	9.7	5.1	0.19	92.3
44°40′N 63°34′W		223	254	56	34		
St. John (Newf.)		35.7	14.6	7.6	7.0	0.29	64.9
47°34′N 52°41′W		210	254	108	77		

TABLE III (continued)

Locality and position		M_2	S_2	K_1	O_1	$\dfrac{K_1 + O_1}{M_2 + S_2}$	$M_2 + S_2 +$ $K_1 + O_1$
Godthaab	H (cm)	135.9	46.9	21.0	9.1	0.16	212.9
64°11′N 51°45′W	$\kappa°$	193	229	127	81		
Cape Whitshed		134.8	47.5	46.1	32.2	0.43	260.7
60°28′N 145°55′W		8	44	130	118		
Sitka		109.7	34.6	45.6	27.8	0.51	217.7
57°03′N 135°20′W		4	34	124	108		
Prince Rupert		194.7	63.8	50.7	31.2	0.32	340.4
54°19′N 130°20′W		6	38	129	114		
Victoria		36.8	9.8	63.0	37.2	2.15	146.8
48°26′N 123°23′W		70	85	146	124		
San Francisco		54.3	12.3	36.9	23.0	0.90	126.5
37°48′N 122°27′W		330	334	106	88		
Magdalena Bay		48.5	30.8	24.1	17.1	0.52	120.5
24°38′N 112°09′W		244	253	71	77		
Mazatlan		32.8	22.6	19.6	13.8	0.60	88.8
23°11′N 106°27′W		265	254	72	75		
Balboa		184.6	48.9	13.5	3.6	0.07	250.6
08°57′N 79°34′W		89	145	342	352		
Valparaiso		43.0	14.2	15.2	10.0	0.44	82.4
33°02′N 71°38′W		279	300	330	286		
Puerto Montt		217.0	111.0	6.0	4.0	0.03	338.0
41°28′S 72°57′W		146	54	7	2		
Orange Bay		58.9	9.2	21.5	17.9	0.58	107.5
53°31′S 68°05′W		104	134	36	347		
Bluff		87.1	15.3	1.8	3.4	0.05	107.6
46°36′S 163°20′E		36	50	116	73		
Auckland		116.4	17.7	7.2	1.6	0.07	142.9
36°51′S 174°46′E		206	268	169	145		
Sydney		51.8	13.1	14.7	9.6	0.37	89.2
33°51′S 151°14′E		251	266	121	94		
Brisbane		67.8	18.8	21.2	11.9	0.38	119.7
27°20′S 153°10′E		290	309	174	143		
Cairns		59.7	34.1	26.6	12.4	0.42	132.8
16°55′S 145°47′E		282	245	190	166		
Thursday Island		35.8	34.9	56.1	29.2	1.21	156.0
10°35′S 142°13′E		51	320	197	148		
Finch, New Guinea		6.8	9.6	25.5	7.0	1.98	48.9
06°35′S 147°50′E		75	124	204	272		
Surabaya		44	26	47	27	1.06	144
07°12′S 112°36′E		351	355	318	284		

TABLE III (continued)

Locality and position		M_2	S_2	K_1	O_1	$\dfrac{K_1 + O_1}{M_2 + S_2}$	$\dfrac{M_2 + S_2 +}{K_1 + O_1}$
Makassar	H (cm)	8	11	28	17	2.38	64
05°06'S 119°24'E	$\kappa°$	70	194	300	278		
Pemangkat		26	4	14	16	1.00	60
01°12'S 109°00'E		111	177	54	7		
Singapore		66	27	24	24	0.52	141
01°18'S 103°54'E		300	348	100	53		
Mergui		167.5	89.2	16.0	6.4	0.09	279.1
12°26'N 98°36'E		310	349	334	315		
Padang		35	14	13	8	0.43	70
00°58'S 100°20'E		177	218	277	265		
Darwin		199.9	104.9	58.2	34.7	0.30	397.7
12°28'S 130°51'E		144	193	336	313		
Fremantle		4.8	4.4	19.4	11.3	3.34	39.9
32°03'S 115°45'E		286	292	300	291		
Adelaide		51.8	51.2	25.3	15.9	0.40	144.2
34°51'S 138°30'E		120	181	52	32		
Kerguelen		43.6	24.5	4.4	6.6	0.16	79.1
49°09'S 70°12'E		9	52	289	292		
Durban		54.9	30.9	5.1	1.5	0.08	92.4
29°52'S 31°03'E		106	138	176	330		
Tamatave		20.5	9.5	2.5	4.1	0.22	36.6
18°09'S 49°26'E		55	54	79	64		
Zanzibar		118.6	57.0	18.9	11.0	0.17	205.5
06°10'S 39°11'E		102	140	38	40		
Aden		47.5	20.6	39.7	20.1	0.88	127.9
12°47'N 44°49'E		227	245	35	37		
Karachi		78.2	29.2	39.9	20.1	0.56	167.4
24°48'N 66°58'E		294	323	46	47		
Bombay		122.2	48.3	42.5	20.1	0.37	233.1
18°55'N 72°50'E		331	4	46	49		
Colombo		17.6	11.9	7.3	2.9	0.35	39.7
06°57'N 79°51'E		50	95	33	62		
Madras		33.2	13.9	9.1	2.9	0.25	59.1
13°06'N 80°18'E		240	270	337	325		
Dublat		140.5	64.2	15.1	5.8	0.10	225.6
21°38'N 88°08'E		291	328	352	338		
Rangoon		170.1	61.1	20.8	8.9	0.13	260.9
16°46'N 96°10'E		132	171	34	24		
Formosa		19	5	19	15	1.42	58
25°06'N 121°42'E		295	285	232	208		
Nagasaki		78	34	23	18	0.37	153
32°42'N 129°51'E		230	256	203	189		

TABLE III (continued)

Locality and position		M_2	S_2	K_1	O_1	$\dfrac{K_1 + O_1}{M_2 + S_2}$	$\dfrac{M_2 + S_2 +}{K_1 + O_1}$
Shantung	H (cm)	59	18	22	13	0.45	112
37°36′N 122°12′E	$\kappa°$	315	2	311	267		
Dairen		99	29	27	20	0.37	175
38°54′N 121°36′E		300	248	1	329		
Miyako		28.4	13.1	22.7	18.6	1.00	82.8
39°38′N 141°53′E		114	150	170	154		
Taraku (Kurile Is.)		27	12	27	20	1.21	86
43°35′N 146°21′E		98	139	160	133		
Petropavlovsk		30.8	8.5	37.5	27.7	1.66	104.5
53°01′N 158°38′E		94	162	131	132		
Honolulu (Hawaiian Is.)		16.0	5.1	14.9	8.2	1.09	44.2
21°18′N 157°52′W		106	102	70	58		
Marshall Islands		46.4	26.4	9.0	5.8	0.20	87.8
05°55′N 169°39′E		104	129	241	212		
Solomon Islands		11.3	8.2	20.7	11.3	1.64	51.5
09°11′S 160°13′E		94	140	207	183		
Tahiti		8.9	7.8	1.1	1.5	0.16	19.3
17°45′S 149°22′W		351	20	278	293		
Gambier Islands		27.5	9.4	1.9	1.0	0.08	39.8
23°07′S 134°58′W		86	38	184	276		

the tidal wave and which therefore may develop into amphidromic waves completely changing the original character of the phenomenon.

It has already been pointed out in several connections that the main features of the tidal phenomenon are its semi-diurnal characteristics. However, in numerous localities, especially in marginal seas, it may be noted that one of the two maximum elevations during a lunar day is much higher than the other one. This phenomenon is due to the presence of more pronounced diurnal constituents and is referred to as the diurnal inequality. In these cases it is usual to speak of a 'mixed' tide. There are also a number of localities where only one high and one low water are observed during a lunar day. This feature is, of course, the consequence of the predominance of the diurnal constituents. A schematic illustration of the interaction between the semi-diurnal and the diurnal tide is given in Fig. 4.

The character of the tide is determined in the following way. The ratio:

$$F = (K_1 + O_1)/(M_2 + S_2)$$

where the symbols stand for the amplitudes of the concerned tidal constituents, is computed. If this ratio is less than 0.25 the tide is characterized as semi-diurnal. The tide is considered to be mixed, but predominantly semi-diurnal, if the ratio lies between 0.25

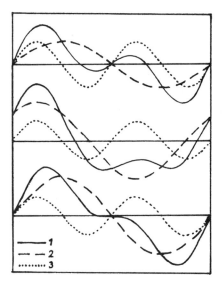

Fig. 4. A schematic illustration of the interaction between the semi-diurnal and the diurnal tides for different amplitudes and differences in phase. Curves (1) are the resultants of the diurnal tides (2) and the semi-diurnal tides (3).

and 1.5. The tide is also said to be mixed, but predominantly diurnal, with the ratio lying between the limits 1.5 and 3.0. Finally, the tide is diurnal if the ratio exceeds the value 3.0. In the last case, however, the possibility is not excluded that two high waters and two low waters may occur from time to time during a lunar day.

A look at the next-to-last column in Table III shows that the semi-diurnal type and the predominantly semi-diurnal mixed type of tide are the most common. According to the data given in the table these two types cover roughly 90% of all cases. There are in the table only four cases of diurnal tides, of which three cases represent the Baltic and the fourth refers to Fremantle in Australia.

Sets of tidal curves representing different types have been given, for instance, by Dietrich (1944) and by Duvanin (1956). Duvanin's curves are reproduced in Fig. 5. At the top of the figure there is the tidal curve for Balboa on the Panama Canal. This curve is a good example of the type of tidal phenomenon which is dominated by the semi-diurnal pattern. The curve shows distinctly the difference between spring tide and neap tide and the diurnal inequality, which is, however, only weakly pronounced. In the Pacific Ocean the tide is frequently mixed, but predominantly semi-diurnal, which indicates that this ocean also reacts to the effect of the diurnal tide-generating forces. The curve for the Fraser River in British Columbia is a good example of this tidal type. At Bangkok the tide is also mixed, but predominantly diurnal. Purely diurnal tides are comparatively rare in the oceans and marginal seas. In addition to Fremantle they occur in the northern parts of the Gulf of Mexico, within the Indonesian Archipelago and in some coastal areas of Vietnam. The curve for the island of Hondo represents this tidal type. A weak influence of the semi-diurnal tide may be traced during neap tides.

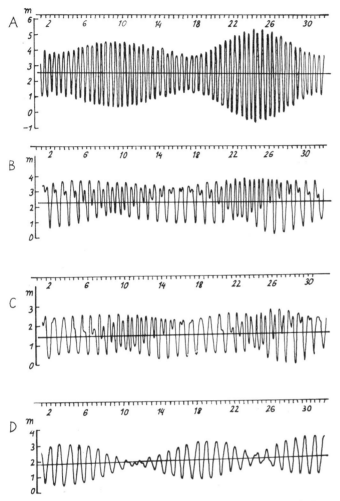

Fig. 5. Curves representing different types of tidal variation during one month. A. Balboa, Panama Canal, B. Fraser River, C. Bangkok, D. Honda Island. (Duvanin, 1956.)

A special comment is necessary concerning the semi-diurnal solar constituent S_2. It has been pointed out by different authors that this constituent, in common with the constituents Sa. Ssa and S_1, is not exclusively gravitational, but is also characterized by a radiational component. According to Cartwright (1968) the radiational tide is mainly caused by the action of the atmospheric tide upon the sea surface, the phase difference — radiational kappa minus gravitational kappa — being about $120°$, since the tidal atmospheric pressure has a minimum 4 hours after the Sun's upper and lower transit. Zetler (1971) has made an attempt to separate the gravitational and radiational components of the constituent S_2, using the data for 31 tidal stations along the west and east coasts of the United States. The above-mentioned phase difference proved to be for the west coast

$133°$ on average and for the east coast $185°$, which thus differs more markedly from the theoretical phase departure. The ratio of the radiational amplitude to the gravitational amplitude, according to the results achieved by Zetler, is 0.16, varying as widely as between the limits of 0.01 and 0.32.

The amplitudes and the phases of the semi-diurnal and the diurnal tidal constituents have generally been proved to be constant for a given observational locality on the coast. The principal features of the tidal phenomenon along the coast could thus be determined. However, it must be kept in mind that the observed results have, as a rule, to be interpolated over irregular coastal areas with complicated shore-lines and frequently highly varying depths, with the consequence that the effect of bottom friction is not always easily taken into consideration. Still more cumbersome is the interpolation of the data representing coastal conditions over the extensive open areas of the oceans with only a few or no observations from the oceanic islands. Electronic computing techniques may be an appropriate approach to this difficult problem, but it must always be remembered that the introduction of a simplified pattern is a generally hazardous approach to numerical computations if exact results are required. Numerical methods based on the knowledge of the tidal elevations and the configuration of the deep ocean do not, as a rule, yield sufficiently accurate results. The practical difficulties are thus considerable.

Rossiter (1963) has summarized the factors necessary for the solution of the problem on a larger scale in the following way. At least six different factors must be taken into account:

(1) the tide-generation force;
(2) the Coriolis force, i.e., the deflecting force of the Earth's rotation;
(3) the satisfaction of boundary conditions;
(4) the dissipation of tidal energy due to the effects of bottom friction and viscosity;
(5) the spatial distribution of depth;
(6) the spatial distribution of water density.

These are the first-order factors. In addition, there remain numerous second-order factors which are not accounted for.

The dynamic equations to be solved for the oceans are:

$$\frac{\partial u}{\partial t} - 2\omega\, v \cos\theta = -\frac{g}{a}\frac{\partial(\xi - \bar{\xi})}{\partial\theta} \tag{5}$$

$$\frac{\partial v}{\partial t} + 2\,\omega\, u \cos\theta = -\frac{g}{a \sin\theta}\frac{\partial(\xi - \bar{\xi})}{\partial\phi} \tag{6}$$

$$\frac{1}{a \sin\theta}\left[\frac{\partial(hu \sin\theta)}{\partial\theta} + \frac{\partial(hv)}{\partial\phi}\right] = -\frac{\partial\xi}{\partial t} \tag{7}$$

In these equations θ is the co-latitude, ϕ the longitude measured positive to the east from a given meridian, t denotes time and u and v are the components of velocity in the direction of increasing θ and ϕ respectively. ξ is the elevation of the water surface above the undisturbed sea level at time t and $\bar{\xi}$ is the equilibrium elevation. h is the average depth, a the Earth's radius, g the acceleration of gravity and ω the angular speed of the rotation of the Earth ($\omega = 2\pi/86,164$ sec $= 7.29 \cdot 10^{-5}$ sec^{-1}). These equations cover the factors (1), (2) and (5) listed above. In the deep oceans the dissipation of the tidal energy is a matter of conjecture.

It is possible that accurate numerical solutions of the above equations will never be obtained. The shapes of the ocean basins may be too complicated for this task. However, there are already available solutions based on the iteration method for the Atlantic Ocean and for a number of marginal seas bordering the Atlantic. Charts representing such surveys for a given tidal constituent, for example M_2 or K_1, contain the corresponding co-tidal lines, which join all the points for which high water of the constituent concerned is observed at the same time. The distribution of the amplitudes in the oceans, represented by the tidal co-range lines characteristic of a given amplitude, is a task which is still more difficult.

Nevertheless efforts are continuously being made to improve the knowledge of the pelagic, i.e., deep-sea, tides. Pressure gauges of different types were constructed in order to obtain direct observations (Eyriès, 1968; Filloux, 1968; Nowroozi et al., 1968; Snodgrass, 1968), while the overall objective of the tidal programs in the deep open oceans was presented, for instance, by Munk and Zetler (1967), Cartwright (1969) and Cartwright et al. (1969). Off-shore measurements of tides across a previously proposed amphidromic area between San Diego (California) and the Hawaiian Islands have been confirmed by Irish et al. (1971). All these results are highly encouraging, but the final solution is still far away.

In order to give a conception of the results achieved by different approaches to the problem, the co-tidal lines for the Atlantic Ocean determined according to the older and the newer method have been reproduced in Fig. 6 and 7. Fig. 6 represents the co-tidal lines for the Atlantic Ocean for the two semi-diurnal tides M_2 and S_2 taken together and referred to Greenwich, according to the results of Von Sterneck (1920). Fig. 7 gives, according to Hansen (1949, 1952a) the theoretical tide in the basin of the Atlantic Ocean. The co-tidal lines are referred to the Moon's transit through the Greenwich meridian and the co-range lines represent the semi-diurnal tidal constituent M_2 in centimetres. It may easily be established that there is a fairly good agreement between the results reproduced in Fig. 6 and 7.

With a few exceptions, the semi-diurnal and diurnal tidal constituents are characterized by rotational waves. This feature is the consequence of the Coriolis force and friction. In the northern hemisphere this rotation is generally anticlockwise, as seen in Fig. 6 and 7, in the southern hemisphere clockwise. In most cases the shape of the rotating wave does not present a completely symmetrical pattern. Generally the rotating waves represent amphidromic regions around an amphidromic point. At some localities there may occur a marked crowding of co-tidal lines and, in extreme cases, the co-tidal lines may be transformed into nodal lines.

The distribution of the tidal range has a close connection with the above-mentioned

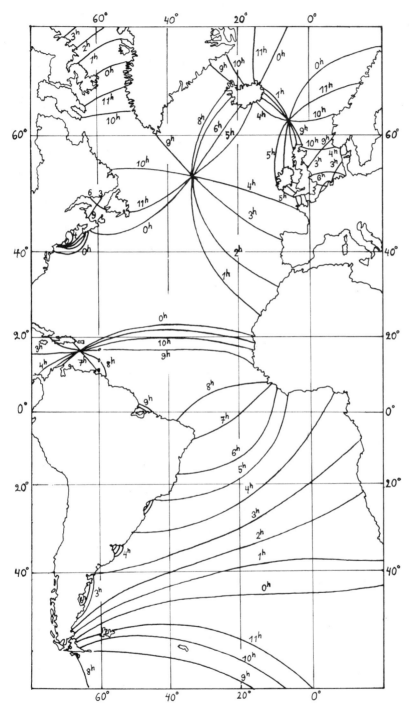

Fig. 6. Co-tidal lines for the Atlantic Ocean representing the added effect of the semi-diurnal constituents M_2 and S_2 (Von Sterneck, 1920).

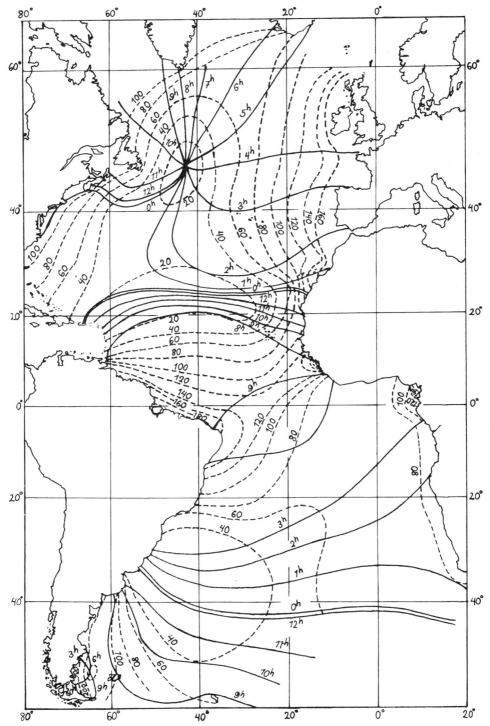

Fig. 7. Co-tidal and co-range lines for the semi-diurnal lunar constituent M₂ in the Atlantic Ocean. Solid lines = time differences of the high water from the Moon's passage through the Greenwich meridian, in hours. Dashed lines = tidal range, in cm (Hansen, 1952a).

crowding of the co-tidal lines. In the centres of the amphidromic regions and along the nodal lines the range is practically zero. Progressing outward from these points and lines, the increase in range is apparent. The slight tidal ranges in the vicinity of the amphidromic points are evident, if these are situated relatively close to an island on which tidal observations have been made. However, the centres of the amphidromes are generally situated in the open parts of the oceans, with the consequence that the most pronounced tidal ranges occur along the coasts of the continents. As to the nodal lines, they may also reach the continental coasts, and in such localities the tidal range will not be marked. It may, for instance, be mentioned that at the coast of southern Brazil the spring-tide range does not exceed 16 cm.

The tide-generating forces also influence water bodies of very restricted dimensions, causing an oscillatory motion, although it may be fairly difficult to measure this in some cases. A tidal oscillation with the range of a few millimetres has, for instance, been established in the Lake of Geneva. The tides observed in gulfs, marginal and Mediterranean-type seas and other more-or-less enclosed water basins are, as a rule, a co-oscillation with the tides in the oceans and not the consequence of the direct effect of the tide-generating forces upon the water surface. Narrow sounds which connect seas like the Mediterranean and the Baltic with the oceans or adjacent seas have, however, a highly moderating effect on the tidal co-oscillation. The equations giving the tidal motion in marginal seas are defined by the following expressions (Rossiter, 1963):

$$\frac{\partial u}{\partial t} - 2\,\omega v \cos\theta \quad = -g\frac{\partial \xi}{\partial x} - ku\sqrt{u^2 + v^2} \tag{8}$$

$$\frac{\partial v}{\partial t} + 2\,\omega u \cos\theta \quad = -g\frac{\partial \xi}{\partial y} - kv\sqrt{u^2 + v^2} \tag{9}$$

$$\frac{\partial\,(hu)}{\partial x} + \frac{\partial\,(hv)}{\partial y} + \frac{\partial \xi}{\partial t} = 0 \tag{10}$$

For the symbols used in these equations reference may be made to eq. 5–7. k is the bottom friction parameter. The second term on the right in eq. 8 and 9 is derived from a quadratic law of bottom friction. However, a linear law is applied in some cases.

Owing to the more restricted dimensions of the seas the preparation of co-tidal and co-range charts for these basins has been easier and more successful than for the oceans. In this respect may be mentioned in particular the efforts of Proudman and Doodson (1924a) and of Hansen (1948, 1952b) for the North Sea.

For sea basins characterized by weak tides there are three factors which are significant for the range of the tidal co-oscillation. We have to take into consideration the range of the oceanic tide at the approach to the sea, the period of the free oscillation of the sea and the dimensions of the entrance between the ocean and the sea. Among the seas with weakly developed tides mention may be made of the Mediterranean, with the exception of the northern parts of the Adriatic Sea and the Aegean Sea; the Baltic, including the transition areas of the Danish Strait, the Kattegat and the Skagerak; the Arctic Ocean and the Sea of Japan. In the Mediterranean the narrow Strait of Gibraltar accounts for the weakly developed tides. In the Skagerak and the Kattegat the tidal range of the co-oscillation is not pronounced as a consequence of the slight tidal range at the wide entrance. The range of the tides decreases still more when passing the narrow transition areas of the Belts and the Strait (Öresund) into the Baltic. In some other cases the causes for the weakly developed tidal oscillations are not quite evident.

It may be of interest to pay more attention to a few of the seas for which relatively extensive results are available. For some of these seas the results are principally based on conservative methods; some others have been studied also with the help of more modern technical processes.

The Red Sea has been selected as a representative of the former type of situation. The first tide gauges were erected in the Red Sea during the last decade of the nineteenth century. In the 1920's the Amiraglio Magnaghi Expedition took place, during which in addition to measurements of tidal currents tidal observations were made at eleven different localities covering a time-span of one to six months.

The Red Sea may be considered a good example of a long and narrow channel. The total length of the sea is 1,932 km and the average breadth 280 km. Since the average depth of the sea is 491 m, bottom friction is of less significance in the final results.

Fig. 8 shows the co-tidal lines in the Red Sea. The distribution of these lines indicates that the principal tide is semi-diurnal in character. Somewhat to the south of the central part of the sea there is a well-developed and relatively symmetric amphidromic oscillation. Outside this amphidrome there is a retardation of high water when progressing from the south to the north. The co-range lines for the Red Sea are reproduced in Fig. 9. The average spring-tide range is at its highest in the northern and southern parts of the sea and decreases towards the central regions, where, as shown in Fig. 8, the anticlockwise amphidrome is situated. In the Red Sea proper the largest spring-tide ranges are approximately 0.5 m, but at Perim in the Bab-el-Mandeb, in the middle parts of the Gulf of Suez and in the Gulf of Aqaba these ranges may reach 1 m. In the parts of the Red Sea where the semi-diurnal tide is weak, the tidal phenomenon is predominantly diurnal. For instance, Port Sudan, is such a locality.

The question of the proportion in range between the independent tide produced in the Red Sea itself and the co-oscillation with the Indian Ocean has been studied by different oceanographers. Only a few of the most important and interesting of these results may be quoted. Defant (1919) showed that the tidal range was of the same magnitude for the independent tide and for the co-oscillating tide. These results were in agreement with the relatively limited tidal observations available at that time. However, Defant (1926), on the basis of renewed computations, drew the conclusion that the tides in the Red Sea are essentially the result of a co-oscillation with those in the Gulf of Aden, while the independent tide produces only slight modifications in the phase of the semi-diurnal tide.

Von Sterneck (1927) was able to draw the conclusion that the ratio of the range of the independent tide to that of the co-oscillating tide was on average for the whole Red Sea approximately 1:3 for the tidal constituent M_2 and 1:4 for the S_2 tide.

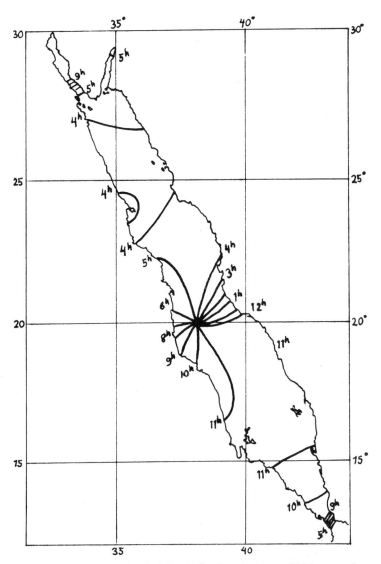

Fig. 8. Co-tidal lines in the Red Sea indicating the time of high water from the Moon's transit at Greenwich (Anonymous, 1963).

These deviating results led to the suggestion that there occurs in the Red Sea an earth tide which must also be taken into consideration. Grace (1930) carried out the computations suggested by Proudman (1928). His recomputations of the semi-diurnal lunar constituent showed that it was not possible to reach a satisfactory agreement between theory and the observed data if the existence of an earth tide was not taken into account. Grace also determined the ratio of the independent tide and the co-oscillating tide. His result was that the ratio was for the largest parts of the Red Sea approximately 3:10. This is in agreement with the ratio determined by Von Sterneck.

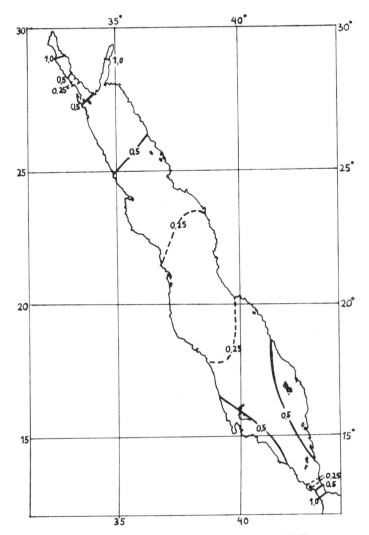

Fig. 9. Co-range lines (in m) in the Red Sea (Anonymous, 1963).

A more comprehensive survey of the character of the tides in the Red Sea was given by Morcos (1970).

A modern approach to tidal problems has been presented by Hansen (1948, 1952b) in the investigations of the tides in the North Sea. The method used by Hansen is called the boundary value method, since it allows the determination of co-tidal and co-range lines for every point in the basin as soon as the tides and tidal currents are known at the boundaries of the area to be examined. Certain simplifications of the motion are necessary. The advective terms and the vertical velocity are neglected. Friction is assumed to be a linear function of velocity. In the numeric work Hansen has replaced the differential

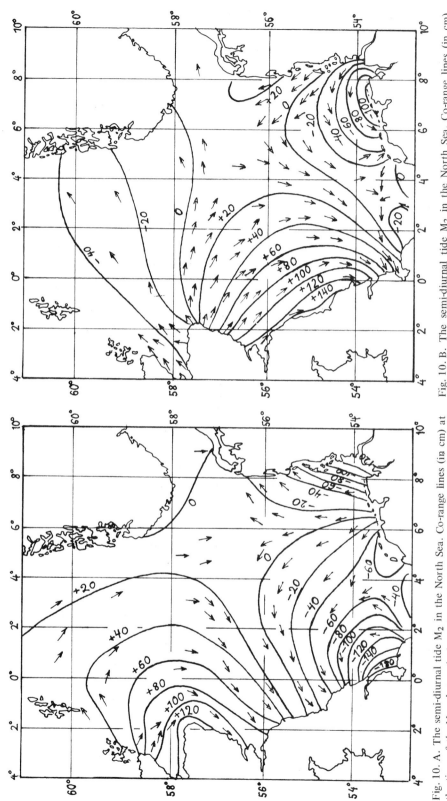

Fig. 10. B. The semi-diurnal tide M₂ in the North Sea. Co-range lines (in cm) at 3 h 6 min after the Moon's transit through the Greenwich meridian. Arrows represent tangents to the lines of equal sea level derived from tidal currents (Hansen, 1948).

Fig. 10. A. The semi-diurnal tide M₂ in the North Sea. Co-range lines (in cm) at the time of the Moon's transit through the Greenwich meridian. Arrows represent tangents to the lines of equal sea level derived from tidal currents (Hansen, 1948).

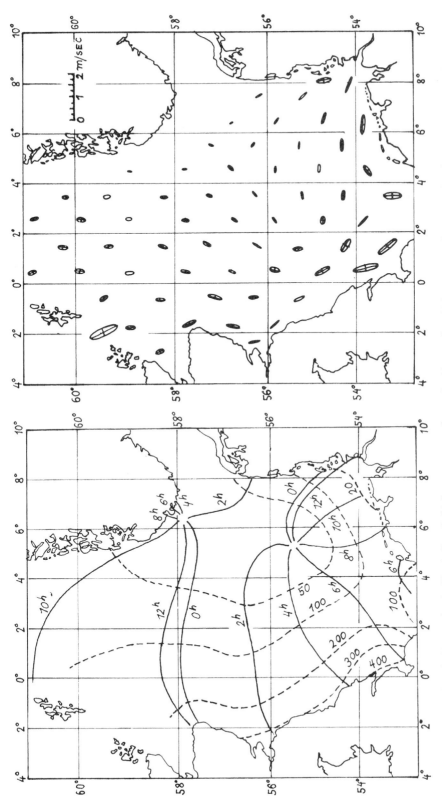

Fig. 10. D. The tidal currents of the M₂ tide in the North Sea. Tidal current ellipses give direction and length of major and minor axes. Scale in the upper right-hand corner refers to the velocity of the current (Hansen, 1948).

Fig. 10. C. The semi-diurnal tide M₂ in the North Sea. Solid lines represent the times of high water (in hours) referred to the transit of the Moon through the Greenwich meridian and dashed lines the tidal range in cm (Hansen, 1948).

equations by a number of linear difference equations. These equations refer to a system of grid points which cover the sea area concerned. At the boundaries of the region the grid points are governed by the observed values. The whole system is based on as many equations as there are grid points. This method allows us to obtain the answer not only to the changes of elevation of the sea level but also of the horizontal motion and thus of the tidal currents.

As mentioned above, Hansen has used this method for the determination of tidal motion in the North Sea (Fig. 10A–D). The results, which refer to the constituent M_2, are based on computations for 21 internal grid points, in addition to the observations along the coasts. If the signs of the co-range lines in Fig. 10A are reversed, the chart represents the sea level for the time point 6 h 12 min after the Moon's passage through the Greenwich meridian. The reversal of the signs in Fig. 10B gives correspondingly the sea level 6 h 12 min later. The two charts thus convey a picture of the changes of the configuration of the sea-level surface during an entire tidal period. Fig. 10C gives the distribution of the tidal range and the time of the occurrence of high water of the M_2 tide. It confirms earlier results for the tides in the North Sea. Fig. 10D shows the tidal current ellipses indicating the direction and length of the major and minor axes.

In principle the boundary value method offers an opportunity to compute the tides for a definite oceanic or sea region, starting from the boundary conditions that the direction of the tidal current in the vicinity of the coast is parallel to the coastal line, but without any knowledge of the actual sea-level distribution at the boundaries. However, in this instance the work involved in the computations becomes very extensive. The calculations must, therefore, as a rule rely upon sea-level observations made along the coasts and on the islands.

It has already been pointed out in several connections that the tidal phenomenon in the oceans and marginal seas deviates to a high degree from the picture given by the equilibrium theory of tides. These deviations increase still more when the tide approaches shallow waters. In deeper waters long-period, diurnal and semi-diurnal tidal constituents have been observed, while species of higher order have not been included in the harmonic system of the tide-generating potential. In shallow water 'over-tides' corresponding to ter-, quarter-, sixth- and eighth-diurnal constituents must frequently be taken into consideration in order to obtain satisfactory results. This fact is very important, and it must not be ignored for the more complete understanding and prediction of the tides in shallow waters.

For the case of a straight, narrow channel or estuary the equations concerned may be written in the form:

$$\frac{\partial u}{\partial t} + u\frac{\partial u}{\partial x} = -g\frac{\partial\,[h+\xi]}{\partial x} - \frac{b\,u|u|}{h+\xi} \tag{11}$$

$$\frac{\partial\,[b\,(h+\xi)\,u]}{\partial x} = -b\,\frac{\partial\,[h+\xi]}{\partial t} \tag{12}$$

In these equations the horizontal axis x is measured along the channel or estuary and u is the average velocity from the bottom to the surface in this direction. t is the time, h the mean depth and b the breadth of the channel. As before ξ is the elevation of the water surface, g refers to the acceleration of gravity and k is a parameter representing bottom friction.

In shallow-water and estuary studies the elevation of the water surface is no longer negligible in comparison with the mean depth h and the product $(h + \xi) \cdot u$ must therefore be introduced in the eq. 11 and 12. It must also be taken into account that the breadth b is generally a function of the distance x. In addition, other factors contribute to distort the shape of the tidal wave entering a shallow-water area. There arise not only odd species of shallow-water constituents, but also combinations of astronomical tides representing higher species. The phenomenon is highly complicated and, in addition, the particular shallow-water regions have their specific characteristics. It is therefore not possible to describe the phenomenon in detail here and a few examples characterizing these species and giving the formulae for their angular speed may therefore be sufficient in order to give a conception of the tidal constituents involved and their possible combinations. These examples are collected in Table IV.

The significance of the shallow-water tides appears distinctly from the analysis performed by Zetler and Cummings (1967) for Anchorage, Alaska, in order to achieve more accurate tidal prediction for this harbour. The recent development of techniques for identifying frequencies typical of the tidal spectrum at a particular locality made it possible to compute the shallow-water tides by harmonic methods. The use of the electronic computer did not restrict the number of constituents and not less than 114 constituents were included by Zetler and Cummings. Of these constituents, approximately 40% had a frequency of 0–2 cycles per day and the remaining constituents covered as high frequencies as 3–12 cycles per day.

So far attention has mainly been directed to the vertical movements of the sea level as a result of the effect of tide-generating forces. The increase and decrease in sea level is

TABLE IV

SOME EXAMPLES OF TIDAL CONSTITUENTS CHARACTERISTIC OF SHALLOW WATER

Symbol[1]	Formula for angular speed
MNS_2	$M_2 + N_2 - S_2$
MSN_2	$M_2 + S_2 - N_2$
MN_4	$M_2 + N_2$
M_4	$2M_2$
SK_4	$S_2 + K_2$
$2MN_6$	$2M_2 + N_2$
M_6	$3M_2$
$4MS_6$	$4M_2 - S_2$
$3MS_8$	$3M_2 + S_2$
M_8	$4M_2$

self-evidently not possible without a horizontal motion of the water back and forth which occurs with the same rhythm as the tides themselves. These horizontal movements of the water are called tidal currents. Owing to the significance of tidal currents, navigation charts representing these currents are given not only in scientific publications but also in nautical handbooks.

Although observations on tidal currents – mainly owing to the greater cost involved in such measurements – are considerably more restricted than those on tidal records, those observations that have been made, show tidal currents to have the same irregularities in time and space as the tides. In some cases they may be still more complicated than the tides, since they are generally two-dimensional. In the open sea current direction varies with current velocity. In estuaries and in the immediate vicinity of the coast only alternating motion is possible.

Owing to the close relationship between the vertical and the horizontal water motion, both being the consequence of the tide-generating forces, the time distribution of the tidal currents is known. Therefore, the same harmonic analysis method as applied to the tides may be used for the tidal currents. Two perpendicular components of the current vectors, i.e., the north and the east component, can be conveniently analyzed. Expressions analogous to eq. 4 are characteristic of the components of the current. Only in quite exceptional cases are the two tidal current components in or out of phase, and the tidal currents are therefore generally rotatory. The direction of the flow rotates clockwise or anticlockwise during the period of a tidal cycle. In the cases where the tide is predominantly semi-diurnal or diurnal the tidal current ellipse corresponds to the cycle concerned. The magnitude and the ratio of the ellipses' axes vary from tide to tide. The numerically determined ellipses, which are presented in Fig. 10D, refer to the M_2 constituent, such as it appears at the centered time, i.e., in between the spring and neap tides. In regions characterized by a mixed tide the features of the tidal current ellipses are more complicated, and have to be examined separately for every individual case.

Even in the cases of relatively regular elliptic tidal currents, the relationship between the phase and the range of the tide and the phase and the velocity of the tidal current may be fairly complicated. This is a consequence of the fact that the characteristics of the tidal currents, are dependent upon the rotation of the Earth, in addition to the effect caused by the tide-generating forces. Moreover, the disturbing contribution of such factors as the depth of the water and the bottom topography is still more accentuated for the tidal currents than for the tides.

If the water in the oceans were homogeneous from the surface down to the bottom and the water pressure at every point in the water column determined by the hydrostatic law, the tidal currents would be uniform with depth. However, the presence of internal waves distorts the conformity between theory and observations. In actual cases nonuniform conditions are characteristic of tidal currents throughout the whole water column. There are considerable deviations in phase and the most pronounced phase jump seems to be associated with the pycnocline, i.e., the layer of the most marked increase in water density.

The velocity of the tidal current may increase as a consequence of the shallowing of the ocean depth towards the shelf or the occurrence of a submarine ridge. The lateral narrowing in straits also has a similar effect. Moreover, hydraulic tidal currents may occur in straits linking areas in which high and low water are more or less out of phase. The extremely strong currents in the Strait of Messina have been known since antiquity and have been examined in detail by Defant (1940). Another example is the Seymour Narrows between the American continent and Vancouver Island, where in the northern part of the narrows tidal currents may reach as high a velocity as 10 knots.

The energy associated with tidal currents may in some cases be enormous. Different projects have therefore been undertaken in order to utilize the energy of these currents. Difficulties have been considerable since changes in energy during a tidal period are fairly large, oscillating from zero to the maximum and vice versa. However, in the Gulf of St. Malo, where the tidal range reaches one of the highest values in Europe, a tidal current power station is in operation.

A significant factor in the studies of tidal currents is friction. This friction occurs along the bottom of the oceans and seas, along the coasts and, in polar regions, is also effective at the ice cover. According to Jeffreys (1952) about half of the total tidal energy present in the oceans is dissipated each day owing to the effect of friction. Because of bottom friction the velocity of the tidal currents is nil at the ocean bottom.

The interaction between tidal currents and ice cover in polar regions is only rarely taken into account in tidal investigations. Nevertheless, a part of the tidal energy is continuously working on the deformation of the ice cover, the breaking up and the hummocking of the ice. Also it must be remembered that while the ice cover acts to diminish the amplitude of the tide, the friction between the water masses and the lower side of the ice cover lessens the velocity of the tidal currents.

Recently the possibility of a spacecraft tide programme (Zetler and Maul, 1971) has been envisaged. So far it is too early to give the details of this programme, although it seems that useful values may be reached by this method, at least for the most pronounced tidal constituents such as M_2 and K_1. According to Zetler and Maul a further improvement in accuracy of the results may be anticipated, since similar results will be reached in a number of adjacent and overlapping areas and therefore at least square contouring studies for each important tide will produce co-tidal and co-range charts which presumably will be more accurate than the individual values for the separate areas. Although the computations will require extensive work, they would seem to be quite feasible.

Long-period tides

While the semi-diurnal and diurnal tidal constituents have been subjected to comprehensive studies by different scientists, the long-period tides have been treated rather scantily. Nevertheless, the contribution of these constituents to the tidal phenomenon as a whole is by no means negligible. According to Doodson (1921) there are not less than 99 long-period tidal constituents, the number of the diurnal tidal constituents being 157 and of the semi-diurnal 129. A large number of the long-period tidal constituents have, however, no practical significance, their relative coefficients being less than one percent of that of the most pronounced tidal wave, i.e., the semi-diurnal constituent M_2. In Table II there are therefore reproduced only ten long-period constituents, since their relative coefficients all exceed the above-mentioned limit. The added effect of these ten constituents is somewhat higher than half the effect of M_2. In order to get a conception of the different types of long-period tidal constituents, they have been divided by Maximov (1970, p. 51) into representative groups and the relative percentage of the potential of these groups computed. The results achieved by Maximov are the following:

Constituents covering several years — group M_N 10.71%
Annual and semi-annual constituents — group Ssa   14.24%
Monthly constituents — group Mm   22.31%
Semi-monthly constituents — group Mf   41.11%
Ter- and quarter-monthly constituents — group Mt   11.63%

Moreover, it may be mentioned that the basic term of the group Mf (075.555 according to Table II) corresponds to 25.10% of the total potential of the long-period tidal constituents and the basic term of the group Mm (065.455) to 13.25% of this potential.

The computations of the long-period tidal constituents in the oceans are based on the following equations representing the equilibrium tide:

$$W = V (1 - 3 \sin^2 \varphi) \cos \psi$$

$$\Delta H = \frac{W}{g}(1 + k - h)$$

In these equations W is the potential of the constituent, V the relative coefficient, ψ the argument, φ the geographical latitude, g the acceleration of the Earth's gravity, ΔH the disturbance in mean sea level caused by the constituent concerned according to the equilibrium theory, and $(l + k - h)$ is a factor representing the resilient and elastic properties of the Earth's yielding crust. The most usual estimate is $k = 0.27, h = 0.60$, and thus $(l + k - h) = 0.67$.

According to Maximov (1970) the equilibrium equations for the ten long-period tidal constituents listed in Table II are as follows:

W_{M_N} = $-857.0 (1 - 3 \sin^2 \varphi) \cos N'$ (19-year period)

W_{Sa} = $153.8 (1 - 3 \sin^2 \varphi) \cos (h-p_s)$ (annual)

W_{Ssa} = $953.1 (1 - 3 \sin^2 \varphi) \cos 2h$ (semi-annual)

W_{MSm} = $206.4 (1 - 3 \sin^2 \varphi) \cos (s-2h+p)$ (monthly)

W_{Mm} = $1079.6 (1 - 3 \sin^2 \varphi) \cos (s-p)$ (monthly)

W_{MSf} = $179.2 (1 - 3 \sin^2 \varphi) \cos (2s-2h)$ (semi-monthly)

$$W_{Mf} = 2046.0 \, (1 - 3 \sin^2 \varphi) \cos 2s \qquad \text{(semi-monthly)}$$

$$W_{f_N} = 841.7 \, (1 - 3 \sin^2 \varphi) \cos (2s + N') \qquad \text{(semi-monthly)}$$

$$W_{Mt} = 391.8 \, (1 - 3 \sin^2 \varphi) \cos (3s - p) \qquad \text{(ter-monthly)}$$

$$W_{Mtt} = 162.3 \, (1 - 3 \sin^2 \varphi) \cos (3s - p + N') \quad \text{(ter-monthly)}$$

Starting with the first of the above equations it must be mentioned that the declination of the Moon, being on average $23°27'$, varies during the period of 18.61 years within the limits of $28°35'$ and $18°19'$. It is just this variation which is represented by the MN tidal constituent, generally called the 19-year constituent or the nodal tide.

The potential of the nodal tide is one of the more marked ones in the group of long-period tidal constituents. It is therefore by no means surprising that the effect of this tide upon different oceanographic and meteorological phenomena has been distinctly observed. This effect has thus been established not only with regard to the variation in sea level, but also in the changes of water temperature and water circulation in the oceans and in the fluctuations of atmospheric pressure and the general features of atmospheric circulation.

The above equation for the 19-year tidal constituent results in the following latitudinal distribution of the maximum sea-level heights:

Lat. (N and S)	0°	10°	20°	30°	40°	50°	60°	70°	80°	90°
ΔH_{M_N} (mm)	5.9	5.3	3.8	1.4	−1.4	−4.5	−7.3	−9.7	−11.2	−11.7

These data indicate that there is a maximum at the equator, the nodal line running approximately along the latitudes of $35°N$ and $35°S$, while the poles are characterized by a maximum which is twice as large as that at the equator. These features being typical also of the other long-period tidal constituents; these tides are frequently referred to as the circumpolar tides. Therefore, they have been investigated more assiduously for the polar regions, most of the studies covering the Arctic Ocean.

Although the theoretical basis of the nodal tide is firmly established and the mathematical analysis by Proudman (1960) clearly indicates that this tidal constituent should follow the equilibrium law, it is evident that the problem of the nodal tide in the oceans and seas is by no means solved. The study of the problem is also obstructed by the regrettable fact that the number of tidal stations with reliable data covering at least a period of 19 years, but preferably 37 years, is at the present time fairly restricted.

The investigations concerned with the nodal tide have followed two principal directions: either an effort has been made to determine the amplitudes and phases of this tide in different parts of the world oceans, in order to reconstruct the character of the world-wide distribution of the tidal constituent and its relationship to the equilibrium tide, or the principal purpose of the researches has been the elimination of the effect of the nodal tide from the data concerned before the determination of the mean sea level. The latter objective will be considered later, when dealing with this problem, while a few words must be devoted here to the first of these directions.

TABLE V

THE HARMONIC CONSTANTS OF THE NODAL TIDE ACCORDING TO THE RESULTS
ACHIEVED BY DIFFERENT AUTHORS

Author	Amplitude (mm)	Phase (years)
Rossiter	16.5	13.6
Lisitzin	19.5	10.9
Maximov (Atlantic Ocean)	26.6	11.6
Maximov (Baltic Sea)	23.1	10.5
Average	21.4	11.6

One of the first attempts, at least since the time of Darwin, to obtain a global picture of the distribution of the 19-year tidal constituent over the oceans was made by Maximov (1954), in a study of this phenomenon by means of periodograms. The author did not consider the lunar period as strictly constant, but as varying for the particular stations within the limits of 17.8 and 21.2 years. In spite of this fact Maximov was able to show that there seemed to be a difference of approximately 140° between the average phase lags in the Atlantic Ocean on the one hand, and in the Pacific Ocean on the other hand. Also Rossiter (1954a) has made an attempt to compute the harmonic constants of the 18.6-year cycle for the North Sea, basing his results on four tidal stations. Lisitzin (1957d), on the basis of data for 27 tidal stations covering all oceans, has been able to conclude that the differences in phase lag for stations situated approximately at the same latitudes, but at considerable distances from each other, is of the same magnitude as the difference in longitudes. Maximov (1959) not only continued the investigations on the nodal tide, but has, in addition, made an attempt to compile and compare all available older results. This author based his computations on 17 stations in the Atlantic Ocean and 27 stations in the Baltic Sea. The average harmonic constants of the nodal tide determined on the basis of the results of different scientists are collected in Table V.

Finally, Maximov (1970, p.88) computed the following equations for the 18.6-year tidal constituent utilizing the particular results:

Rossiter, 4 stations in the North Sea

$$\Delta H = 15.7 \, (1 - 3 \sin^2 \varphi) \cos (N' - 83°) \, \text{mm}$$

Lisitzin, 11 stations in the North Atlantic and the North Pacific Ocean

$$\Delta H = 21.8 \, (1 - 3 \sin^2 \varphi) \cos(N' - 75°) \, \text{mm}$$

Maximov, 17 stations in the North Atlantic Ocean

$$\Delta H = 31.6 \, (1 - 3 \sin^2 \varphi) \cos (N' - 83°) \, \text{mm}$$

Maximov, 27 stations in the Baltic Sea

$$\Delta H = 20.1 \, (1 - 3 \sin^2 \varphi) \cos (N' - 23°) \, \text{mm}$$

As a general conclusion, it may be mentioned that, in spite of the fact that the above results are based on different stations and deviating methods of computation, they show a considerable conformity. The amplitude of the tide is by no means negligible, since in higher latitudes the constituent brings about changes in the mean sea level of the magnitude of 40–50 mm. Extrapolating the results for the latitudes 50°N and 60°N to cover the whole Earth, Maximov (1959) determined the following values for the amplitudes referring to the particular latitudes:

Lat. (N and S)	0°	10°	20°	30°	40°	50°	60°	70°	80°	90°
ΔH (mm)	22	20	14	5	−5	−17	−27	−36	−41	−44

Rossiter (1967) tried to prove the existence of the equilibrium nodal tide in the oceans. Since some of the available sea-level data covering a sufficiently extensive period were rather unsubstantial, an average value of the amplitudes and phase lags was computed by Rossiter for particular European countries. By this procedure the scattering of the values was reduced by a certain amount. The more substantial data seem to give some weak support for the existence of an equilibrium nodal tide in the oceans. Concerning the amplitudes there seems to be only slight evidence for the dependence upon the latitudes required by the equilibrium theory. Rossiter himself has pointed out that additional proof covering more extensive oceanic regions is necessary in order to achieve a more convincing picture of the entire phenomenon.

The second long-period tidal constituent mentioned in Table II is the annual solar tide, Sa. In spite of the fact that this constituent has a fully testified astronomical background, the seasonal meteorological and hydrographic contribution to its range is highly accentuated and, especially when taking into consideration the significance of the annual variation in sea-level research, it seemed to be more appropriate to deal with this constituent in a separate chapter (pp. 109–128). In order to give a conception of the latitudinal distribution of the magnitude of the Sa-constituent in accordance with the equilibrium theory, the values concerned are reproduced below:

Lat. (N and S)	0°	10°	20°	30°	40°	50°	60°	70°	80°	90°
ΔH_{Sa} (mm)	1.1	1.0	0.7	0.3	−0.3	−0.8	−1.3	−1.7	−2.0	−2.1

The astronomical contribution of the Sa-tidal constituent is thus, according to the equilibrium theory, at the utmost about 4 mm.

On the contrary, the semi-annual tidal constituent Ssa is, in terms of the strength of its potential, the third in order among the members in the group of long-period tidal constituents, next only to the semi-monthly tide Mf and the monthly tide Mm. The equilibrium theory gives the following distribution of the maximum amplitudes of the Ssa-tide with latitude:

Lat. (N and S)	0°	10°	20°	30°	40°	50°	60°	70°	80°	90°
ΔH_{Ssa} (mm)	6.5	5.9	4.2	1.6	−1.6	−5.0	−8.1	−10.7	−12.4	−13.0

Owing to the considerable amount of sea-level data, published by the International Association of Physical Oceanography (Association d'Oceanographie Physique, 1940, 1950, 1953, 1958, 1959, 1963, 1968) giving the average monthly and annual values, it has been possible to analyze these data harmonically with the principal purpose of obtaining a conception of the global distribution and of the characteristics of the semi-annual tidal constituent in the oceans. The relevant determinations have been performed by Polli (1942), Lisitzin (1956), Maximov (1959, 1965b,c) Maximov and Smirnov (1965), Karklin (1967) and others. The greatest work in systematizing the results achieved by different oceanographers has been made by Maximov (1970).

Polli was the very first to make an attempt to determine the features typical of the Ssa-constituent in the Atlantic Ocean. Lisitzin has based her investigations on a considerable number of sea-level data (referring altogether to 228 localities) covering the Atlantic, the Pacific and the Indian Oceans and some adjacent seas. Nevertheless, the global coverage was even in this case by no means complete, since sea-level data have not been available for extensive parts of the continental coasts, especially in the southern hemisphere. In addition, oceanic islands were, as a rule, only scarcely represented. The results are reproduced in the chart of Fig. 11. This chart shows the months corresponding to the maximum sea level of the constituent Ssa. The scattering of the data seems to be quite pronounced in some parts of the world oceans. However, Maximov (1965b) was able by compiling the results already available and by adding complementary data, to obtain the general outlines of the features characteristic of the pertinent tidal constituent. The results are reproduced in Table VI.

The phase angles in Table VI distinctly indicate that the solar semi-annual constituent Ssa represents a standing wave in the world oceans. The positions of the nodal lines cannot be determined exactly, but they are assuredly situated in the lower middle latitudes. This is evident from the opposite character of the phase angles in the high and middle latitudes, on the one hand, and of those in the equatorial and adjacent regions, on the other hand.

On average, for all the oceans the following values for the phase angles and phase lags of the Ssa-tide were calculated by Maximov:

	Phase angles	Phase lags
To the north of 40°N	252°	72°
Between 20°N and 20°S	102°	102°
To the south of 40°S	264°	84°

The average value of the phase lag for all the data is 86°, corresponding to a retardation of the semi-annual constituent of 1.4 months.

Concerning the distribution of the amplitudes, it may be mentioned that they should, according to the equilibrium theory, increase to the north from the latitude 35°N and to the south from the latitude 35°S and, in addition, increase, although with the opposite sign of the phase, from the named latitudes towards the equator. Amplitudes, partly levelled and inter- or extrapolated on the basis of the observed tidal data for different latitudes and different oceans, are according to Maximov (1970) as follows on p. 44.

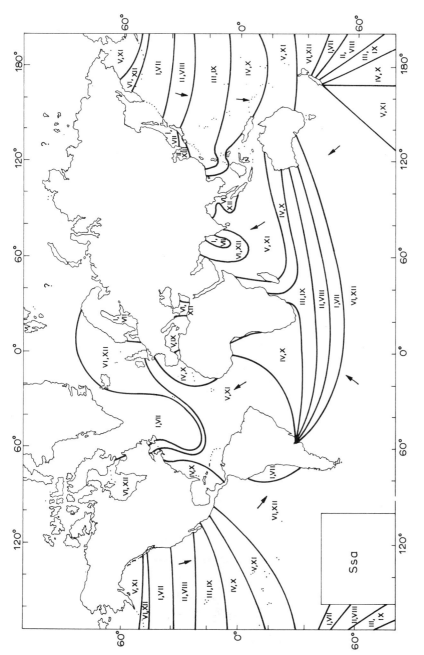

Fig. 11. Co-tidal lines of the semi-annual constituent Ssa. The numbers refer to the months (Lisitzin 1956).

TABLE VI

THE AVERAGE HARMONIC CONSTANTS OF THE SOLAR SEMI-ANNUAL CONSTITUENT Ssa
IN DIFFERENT LATITUDINAL ZONES

Latitude	Number of stations	Amplitude (mm)	Phase angle (°)	Phase lag (°)
Atlantic Ocean				
– 60°N	18	53	228	48
60°N – 40°N	63	38	258	78
40°N – 20°N	25	52	66	66
20°N – 0°	4	45	90	90
0° – 20°S	–	–	–	–
20°S – 40°S	7	36	246	66
Indian Ocean				
40°N – 20°N	16	49	150	150
20°N – 0°	29	58	102	102
0° – 20°S	3	56	90	90
Pacific Ocean				
60°N – 40°N	19	46	264	84
40°N – 20°N	28	42	312	132
20°N – 0°	5	72	114	114
0° – 20°S	3	22	114	114
20°S – 40°S	9	36	300	120
40°S – 60°S	5	32	264	84

Lat.	80°S	60°S	40°S	20°S	0°	20°N	40°N	60°N	80°N
ΔH_{obs} (mm)	−71	−51	−34	2	56	39	−11	−47	−61
$\dfrac{\Delta H_{obs}}{\Delta H_{theor}}$		5.9	6.4	(17.0)	(0.5)	8.0	(9.7)	5.5	5.9	5.1

If the ratios between the observed amplitudes and the amplitudes computed according
to the equilibrium theory for the latitudes 40°S, 20°S and 20°N, for which the number
of tidal stations and thus also of available data are very restricted, are left out of consideration, a more-or-less pronounced relationship between the two series of data may be
noted. In spite of the fact that the average value of the ratio is comparatively high, i.e.,
6.1, this fact may not be considered as evidence against the tidal character of the semi-
annual constituent. In this respect it may be appropriate to point out that fluctuations
characterized by a distinct semi-annual rhythm have been observed in connection with a
considerable number of different phenomena in the oceans. It could, for instance, be
established that this rhythm determines the drift of the ice in the Arctic Ocean and the
general circulation of the water masses and ice drift in the southern oceans. Fedorov
(1959) paid special attention to the effect of the semi-annual tidal constituent Ssa upon
the variations of the intensity of the Gulf Stream and the Kuroshio. Fedorov based his

results partly on direct observations of the velocity and water transport of the two named current systems (Iselin, 1940; Fuglister, 1951; Masuzawa, 1954), and partly on the variation in the difference in sea-level heights between two tidal stations located at the opposite sides of the current (Lisitzin, 1956).

In Table II are listed two monthly and three semi-monthly tidal constituents (MSm, Mm, respectively MSf, Mf and Mf_N). These constituents, although deviating comparatively in their relative amplitudes, may nevertheless be considered jointly.

The equilibrium theory gives the following maximum amplitudes for the five constituents in question:

Latitude (N and S)	0°	10°	20°	30°	40°	50°	60°	70°	80°	90°
ΔH_{Mm} (mm)	7.4	6.7	4.8	1.8	−1.8	−5.6	−9.2	−12.2	−14.1	−14.7
ΔH_{MSm} (mm)	1.4	1.3	0.9	0.4	−0.4	−1.1	−1.8	−2.3	−2.7	−2.8
$\Delta H_{Mm} + \Delta H_{MSm}$	8.8	8.0	5.7	2.2	−2.2	−6.7	−11.0	−14.5	−16.8	−17.5
ΔH_{Mf} (mm)	13.9	12.7	9.1	3.5	−3.4	−10.6	−17.5	−23.0	−26.7	−27.9
ΔH_{Mf_N} (mm)	5.8	5.3	3.8	1.5	−1.4	−4.4	−7.2	−9.5	−11.1	−11.6
ΔH_{MSf} (mm)	1.2	1.1	0.8	0.3	−0.3	−0.9	−1.5	−2.0	−2.3	−2.5
$\Delta H_{Mf} + \Delta H_{Mf_N} + \Delta H_{MSf}$	20.9	19.1	13.7	5.3	−5.1	−15.9	−26.2	−34.5	−40.1	−42.0
ΔH (five const.)	29.7	27.1	19.4	7.5	−7.3	−22.6	−37.2	−49.0	−56.9	−59.5

The above given data indicate that the theoretical range of the equilibrium tide is considerable, being for the monthly constituents in the higher latitudes of the magnitude of 35 mm, and reaching for the semi-monthly constituents in the polar regions almost 85 mm. The total contribution of the monthly and semi-monthly tidal constituents is thus in the high latitudes not less than 120 mm.

TABLE VII

THE AVERAGE HARMONIC CONSTANTS OF THE MONTHLY TIDAL CONSTITUENT Mm IN DIFFERENT LATITUDINAL ZONES

Latitude	Number of cases	Amplitude (mm)	Phase (days)	Mean zonal phase (days)
70°N − 60°N	4	28.1	17.0	
60°N − 50°N	29	43.1	17.0	16.5
50°N − 40°N	24	21.9	15.3	
40°N − 30°N	16	17.1	17.3	
30°N − 20°N	21	22.4	1.3	
20°N − 10°N	19	22.3	0.3	
10°N − 0°	10	10.0	1.6	0.6
0° − 10°S	−	−	−	
10°S − 20°S	−	−	−	
20°S − 30°S	5	20.8	−2.4	
30°S − 40°S	8	27.7	15.1	13.5
40°S − 50°S	8	22.7	11.8	

TABLE VIII

THE AVERAGE HARMONIC CONSTANTS OF THE SEMI-MONTHLY TIDAL CONSTITUENT
MSf IN DIFFERENT LATITUDINAL ZONES

Latitude	Number of cases	Amplitude (mm)	Phase (days)	Mean zonal phase (days)
70°N – 60°N	4	24.4	10.1 ⎫	
60°N – 50°N	29	28.4	11.6 ⎪	9.4
50°N – 40°N	23	21.5	9.4 ⎬	
40°N – 30°N	16	4.3	5.4 ⎭	
30°N – 20°N	20	28.9	1.6 ⎫	
20°N – 10°N	19	22.2	1.5 ⎪	
10°N – 0°	9	9.7	–2.6 ⎬	1.1
0° – 10°S	–	–	– ⎪	
10°S – 20°S	–	–	– ⎪	
20°S – 30°S	4	4.6	5.3 ⎭	
30°S – 40°S	8	25.9	10.6 ⎫	11.2
40°S – 50°S	9	27.7	11.8 ⎭	

It is once again Maximov (1958a, b, 1959, 1960, 1966) who has devoted the greatest interest to the studies of the monthly and semi-monthly tides in the oceans. Indeed, Maximov only has closely examined the three more pronounced constituents of this group, basing his results at least partly on the harmonic analysis of the relevant cycles in sea level made by Schureman (1924). In this way Maximov (1959) achieved the results collected in Tables VII–IX.

The harmonic phase angles in these tables show very distinctly that the three tidal constituents in fact correspond to the stipulations of the equilibrium theory, representing in the oceans a standing wave with the nodal lines situated somewhere in the vicinity of

TABLE IX

THE AVERAGE HARMONIC CONSTANTS OF THE SEMI-MONTHLY TIDAL CONSTITUENT Mf
IN DIFFERENT LATITUDINAL ZONES

Latitude	Number of cases	Amplitude (mm)	Phase (days)	Mean zonal phase (days)
70°N – 60°N	3	34.3	6.2 ⎫	
60°N – 50°N	28	37.7	9.9 ⎪	8.5
50°N – 40°N	25	15.5	8.2 ⎬	
40°N – 30°N	14	14.0	6.5 ⎭	
30°N – 20°N	21	17.0	1.9 ⎫	
20°N – 10°N	20	22.1	0.5 ⎪	
10°N – 0°	8	13.4	0.1 ⎪	
0° – 10°S	–	–	– ⎬	0.8
10°S – 20°S	1	39.6	–1.0 ⎪	
20°S – 30°S	4	14.3	–1.8 ⎭	
30°S – 40°S	8	14.4	5.7 ⎫	
40°S – 50°S	9	22.3	7.6 ⎪	6.8
50°S – 60°S	–	–	– ⎬	
60°S – 70°S	1	56.1	9.0 ⎭	

the latitudes 30°N and 30°S. In the equatorial and adjacent regions the phases of the semi-monthly and monthly constituents correspond approximately to the phase of the potential, while in the middle and higher latitudes the phases are in opposition to the phase of the potential. There is thus a conformity with the equilibrium theory in this respect also. For the tidal constituents of this group there is, on average, a phase lag corresponding to 1.5 solar days.

Maximov (1966) has also given the average weighted values of the amplitudes of the Mm and Mf tidal constituents, covering the Atlantic, Pacific and Indian Oceans. In addition, Maximov has determined the ratios between the observed amplitudes and the amplitudes according to the equilibrium theory. The results are given in the following. They are, for the Mm constituent:

Lat. (N and S)	0–10°	10–20°	20–30°	30–40°	40–50°	50–60°	60–70°	70–80°
ΔH_{obs} (mm)	18.4	22.7	29.7	21.4	21.3	35.2	30.0	29.4
$\Delta H_{obs}/\Delta H_{theor}$	2.6	4.0	9.0	–	5.8	4.8	2.8	2.2

and for the Mf-constituent:

Lat. (N and S)	0–10°	10–20°	20–30°	30–40°	40–50°	50–60°	60–70°	70–80°
ΔH_{obs} (mm)	17.7	18.7	19.0	15.8	19.0	20.3	31.3	30.4
$\Delta H_{obs}/\Delta H_{theor}$	1.3	1.7	3.0	–	2.7	1.4	1.6	1.2

The average value of the ratio for the constituent Mm is 4.46 and for the constituent Mf 1.84. The two series show the same tendency. The highest values of the ratio occur in the temperate latitude, the lowest in the equatorial zone and in the polar regions. However, it is in the higher latitudes that the two constituents show the largest amplitudes, influencing the sea level by 60–70 mm. The results obtained for the Mf tide are graphically presented in Fig. 12.

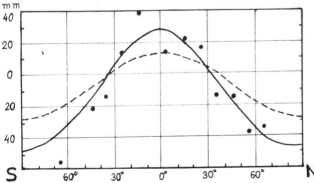

Fig. 12. The latitudinal distribution of the amplitudes of the semi-monthly tidal constituent Mf. Solid line = the amplitudes based on observations, represented by black circles. Dashed line = the amplitudes according to the equilibrium theory (Maximov, 1959).

Table II has already given an indication that the combined effect of the group of the long-period tidal constituents is fairly complicated. To this group belongs the 18.6-year constituent, but also constituents covering only 9 days. According to the equilibrium theory there are also 7-day tidal constituents, although their relative contribution is rather weak, amounting to 2.25% of the total potential of the long-period tidal constituents. As an interesting feature it may, however, be mentioned that the equilibrium theory gives not less than 22 tidal constituents corresponding approximately to periods of one-third or one-quarter of a month.

The 9- and 7-day constituents have been examined very scantily. Nevertheless, Maximov, in his extensive studies of the long-period tidal constituents, paid some attention to this group (Maximov, 1965a; Maximov et al., 1967) basing the results mainly on the data referring to the Arctic Ocean. The maximum amplitudes of the two most pronounced ter-monthly constituents may be given according to the requirements of the equilibrium theory:

Latitude (N and S)	0°	10°	20°	30°	40°	50°	60°	70°	80°	90°
ΔH_{Mt} (mm)	2.7	2.4	1.7	0.7	−0.6	−2.0	−3.3	−4.4	−5.1	−5.4
ΔH_{Mtt} (mm)	1.1	1.0	0.7	0.3	−0.3	−0.8	−1.4	−1.8	−2.1	−2.2
$\Delta H_{Mt} + \Delta H_{Mtt}$ (mm)	3.8	3.4	2.4	1.0	−0.9	−2.8	−4.7	−6.2	−7.2	−7.6

The above values also show that the most marked ter-monthly tidal constituents are weak— even in the case where the two constituents reinforce each other the contributions to the changes in sea level are only of the magnitude of 15 mm. On the other hand, Maximov (1966) was able to show that the total sum of the amplitudes of the 9- and 7-days constituents is 52% of the amplitude of the monthly constituent Mm. One of the principal difficulties in the practical investigations of the ter-monthly and quarter-monthly tidal constituents is, that in order to avoid the effect of the semi-monthly tides, rather lengthy observations are required. For instance, 126 solar days in the case of the 9-day constituent. Also, in utilizing a series of observations over this time-span, one must keep in mind that the main ter-monthly constituent is not exactly 9 days, but 9.1–9.2 days. This fact may, for these weakly pronounced tidal constituents, cause considerable deviations in amplitude and phase. Maximov (1970, p.55) made an attempt to avoid the most marked of these difficulties by studying the occurrence of the 9-day constituents in the Arctic Ocean by means of periodograms. The results were fairly convincing. The amplitude of the main constituent reached 59 mm in the Bay of Tiksy. In more than 75% of all the cases the amplitude exceeded 20 mm, and in 55% of the cases 30 mm. The phase angles showed a considerable scattering, but were on average 228°. This phase angle is an indication that the maximum of the constituent occurs along the coast of the Arctic Ocean, approximately 1.2 days after the occurrence of the minimum of the potential of the tide-generating force. Such a pronounced phase lag shows that the ter-monthly constituent follows the law of a free, not of a forced, wave.

It has already been mentioned that investigations of the 7- and 9-day rhythms are still in their infancy. Nevertheless, it may be assumed that the phenomenon is widespread not only in the oceans but also in the atmosphere. In the oceans, in addition to sea-level changes, the rhythm may be traced in the variations of water temperature, the drift of the ice and the development of internal waves. Since this cycle, in analogy with those of other long-period tidal constituents, is most accentuated in the high latitudes, the phenomena concerned should mainly be examined on the basis of the observational data from the Arctic and Antarctic oceanic regions.

The conditions under which a long-period tidal constituent will follow the equilibrium law was studied more closely by Proudman (1960). Of course, owing to the complexity of the configurations of the oceans, the varying depth and the changing current velocities, this problem can be solved only approximately and separately for the particular oceanic regions. According to Proudman, the long-period tides may maintain their equilibrium character exclusively in the cases where the period of the constituent exceeds a 'critical period', which can be determined for the different regions in the oceans by means of a critical term. This term K was given by Proudman by the formula:

$$K = \frac{1}{100\pi} \cdot \frac{v}{h}$$

where v is the maximum velocity of the long-period tidal current (cm/sec) and h the water depth in metres.

One would expect that the condition for the applicability of the equilibrium theory is that the period of the tidal constituent is long compared with $2\pi/K$. Now:

$$\frac{2\pi}{K} = 200\ \pi^2\ \frac{h}{v}$$

but, self-evidently, the ratio h/v must vary considerably in different parts of the oceans.

It may be of interest to determine the critical period for the deep oceans. Starting from the assumption that the mean depth of the ocean is 4,000 m and the maximum velocity of the current 5 cm/sec, the critical period may be computed to be equal to 5.1 years. For the continental shelf with a depth, let us say, 20 times less, the critical period should be of the magnitude of approximately 100 days. According to these results, only the nodal tide with the period covering 18.61 years would follow the equilibrium law. This result seems, however, to be contradicted by the observed data. The highly complicated shapes of the oceans and seas are probably the main cause of this disagreement between theory and observations.

Before leaving the set of problems connected with the tide-generating forces and the tidal phenomena in the oceans, attention should be paid to the constant term of the tide-generating potential of the Moon and Sun, i.e., the so-called deforming force of these astronomical bodies. The constant term of the tide-generating forces is given by the equations:

$$M_{const.} = 6599.90\ (1 - 3\ \sin^2 \varphi)$$

$$S_{const.} = 3062.17\ (1 - 3\ \sin^2 \varphi)$$

By summing up these equations we obtain:

$$M_{const.} + S_{const.} = 9662.1\,(1 - 3\sin^2\varphi)$$

The consequence of the deforming forces is that the mean sea level in the world oceans is at the equator and the low latitudes higher, and in the temperate and high latitudes lower, than could be expected if the mean sea level over the whole of the globe corresponded to the geoid. According to the equilibrium theory this deformation of the surface level of the oceans corresponds to the following values:

Latitude (N and S)	0°	10°	20°	30°	40°	50°	60°	70°	80°	90°
$\Delta H_{M const.} +$										
$+ \Delta H_{S const.}$ (mm)	66	60	41	16	−16	−49	−83	−109	−126	−133

These values show that the difference in sea level between the equator and the poles should be 20 cm. It is rather difficult to decide whether these features correspond to the actual conditions in the oceans. The main difficulty consists in the fact that not only the tide-generating forces, but also the uneven distribution of these forces (Maximov, 1967) and, in addition, meteorological and hydrographic factors, cause deformation of the water surface and thus deviations between this surface and the geoid. Once again it was Maximov (1966) who made an attempt to solve the problem by assuming that the ratio between the actual deviations and those required by the equilibrium theory is of the same magnitude as for the long-period tidal constituents. Supposing that this ratio is, on average, 3.7, Maximov achieved the following results for the estimated deformation:

Latitude (N and S)	0°	10°	20°	30°	40°	50°	60°	70°	80°	90°
$\Delta H_{M const.} +$										
$+ \Delta H_{S const.}$ (mm)	245	223	150	61	−63	−180	−306	−402	−468	−491

In this case the deformation of the surface of the oceans should be highly pronounced owing to the influence of the constant deforming forces of Moon and Sun. The height difference in sea level between the equator and the poles should amount in this system to 74 cm. Such a difference in sea level must without doubt exercise a considerable effect upon the water circulation in the oceans. This effect might possibly be traced in the transport of surface water masses from the equator towards the higher latitudes and in the compensation transport of the deep water in the opposite direction.

In this respect it must also be taken into consideration that the differences in sea-level heights given above refer only to the constant term of the tide-generating forces. Adding or subtracting the fairly complicated effect of the long-period tidal constituents, the sea-level gradient will show notable variations. Maximov (1966) estimated the total contribution of the long-period tidal constituents in the most favourable cases, finding the following:

Latitude (N and S)	0°	10°	20°	30°	40°	50°	60°	70°	80°	90°
$\Delta H_{long period}$ (mm)	58	53	37	15	−15	−43	−73	−95	−111	−116

Adding and subtracting these results from the data representing (according to Maximov) the constant deformation we find:

Latitudr (N and S)	0°	10°	20°	30°	40°	50°	60°	70°	80°	90°
Maximum (mm)										
$\Delta H_{const.} + \Delta H_{long\ per.}$	303	276	187	76	−78	−223	−397	−497	−579	−607
Minimum (mm)										
$\Delta H_{const.} - \Delta H_{long\ per.}$	187	170	113	46	−48	−137	−233	−307	−357	−375

The sea-level differences in the oceans between the equator and the poles must then, according to the above speculations, vary within the limits of 56 and 91 cm. Whether these computations have a real background can hardly be decided at the present time.

A short survey of long-period tidal phenomena has recently been given by Maximov et al. (1972).

THE CHANDLER EFFECT – CHANGES IN THE ROTATION OF THE EARTH

The Chandler effect is understood to mean variations in sea-level arising in the oceans and seas as the result of the variable force which is the consequence of the movement of the instantaneous axis of the Earth's rotation. The term 'pole tide' for this effect was introduced by Darwin (1898), but this designation is not quite correct, since the variations in sea level are caused by the changes in the centrifugal force of the Earth in its orbital motion around the centre of the mass of the Earth—Moon system bringing about changes in the movement of the instantaneous axis of the Earth. The pole tide is thus in the proper sense of the word not an astronomical phenomenon, its origin being connected with processes on the Earth. However, the term pole tide is so firmly established in scientific literature dealing with sea level and its fluctuations that there seems to be little reason for discriminating against it. The term Chandler effect is, however, used alongside the term pole tide and preference should always be given to the former designation. It was S. Chandler who discovered in 1892 that there were changes in latitude characterized by the period of 14 months (427 days) and that these variations were caused by the free oscillation of the Earth's poles (Chandler motion).

Darwin indicated that the pole tide should move around the pole in the direction of the Chandler motion, i.e., from west to east. In addition, Darwin noted that the variation in mean sea level brought about by the Chandler effect was in character related to the long-period tidal constituents proceeding in the oceans as a forced wave.

The equilibrium theory of the pole tide was developed by Schweydar (1916). The equations are based on the following considerations:

The potential V of the centrifugal force of the Earth's rotation is given by the equation:

$$V = -\tfrac{1}{2}\,\omega^2\,a^2\,\sin^2\theta$$

where ω is the angular speed of the Earth's rotation, a the average radius of the Earth and θ the co-latitude.

Owing to the Chandler motion the magnitude and direction of the centrifugal force change continuously at every point on the Earth. The potential of the deforming force is thus:

$$W_p = \Delta V = -\tfrac{1}{2}\,\omega^2 a^2\,2 \sin\theta\,\cos\theta\,\Delta\theta$$

$\Delta\theta$ corresponds to the radius vector of the movement of the poles and may be expressed by:

$$\Delta\theta = x \cos\lambda + y \sin\lambda$$

where x and y are the rectangular coordinates of the instantaneous poles in relation to their average position and λ the longitude of the point under consideration. Thus one obtains the expression:

$$\Delta V = -\tfrac{1}{2}\,\omega^2 a^2\,(x \sin\lambda + y \sin\lambda)\,\sin 2\,\theta$$

The deformity caused by the Chandler effect may therefore, in complete conformity with the deformation caused by the tide-generating forces (p. 38) be given by:

$$\Delta H = \frac{\Delta V}{g}\,(1 + k - h)$$

This equation makes it possible, starting with Greenwich longitude ($\lambda = 0$), to compute the maximum amplitudes of the pole tide for different latitudes. The first data for these amplitudes were given already by Schweydar. Maximov et al. (1970) recomputed these data using more recent determinations of the values for k and h, and assuming that the x-coordinate of the pole is equal to 0.22 sec, which corresponds to the results of the harmonic analysis of a series of coordinates during the period 1947—1960. The amplitudes of the equilibrium variations were for the longitude of Greenwich:

Latitude (N and S)	0°	10°	20°	30°	40°	45°	50°	60°	70°	80°	90°
$\Delta H_{Chandler}$ (mm)	0.0	2.8	5.2	7.0	8.0	8.1	8.0	7.0	5.2	2.8	0.0

These data represent an oscillation with the nodes at the equator and the poles and the highest sea level at the latitudes of 45°N and 45°S.

Passing from theory to observation, it may be noted that no further pronounced progress in the studies of the Chandler effect upon the sea level was made before the 1950's. The studies of Nicolini (1950), Baussan (1951), Maximov (1952, 1956a,b), Haubrich and Munk (1959), Shimizu (1963) and Jessen (1964) were in the next developmental stage. Although the approaches of the different scientists to the problem were very varied, the results indicated very distinctly the significance of the pole tide in the studies of sea-level variations on a world-wide scale. Baussan utilized long series of sea-level observations at numerous localities in the Atlantic, Pacific and Indian Oceans and found that the occurrence of a cycle covering 14 months, or thereabouts, could be established everywhere in the oceans. Baussan also was the first to note that the ampli-tudes of the Chandler effect are, in the oceans, dependent upon the latitudes, reaching their highest values as a rule in the middle latitudes of the two hemispheres. Furthermore, Baussan was able to establish that the amplitudes of the pole tide vary considerably with time. The amplitudes could, for instance, vary within such large limits as 3 and 59 mm. Fig. 13 illustrates according to Maximov (1970), the sea-level variations caused by the Chandler effect for the stations at Esbjerg, Buenos Aires and San Francisco.

Maximov must again be considered as the most outstanding name with regard to the studies of the Chandler effect, not only in the oceans but also in the atmosphere. This oceanographer was able to show, in different papers, basing his results on tidal series covering at least seven years and thus corresponding to six 14-month periods, that the distribution of the amplitudes and phases of the Chandler effect upon the sea level in fact proves that the circumpolar asymmetric wave from west to east is caused by the motion of the instantaneous axis of the Earth. The most characteristic features of this wave are that the phases are opposite in the northern and southern parts of the Atlantic and Pacific

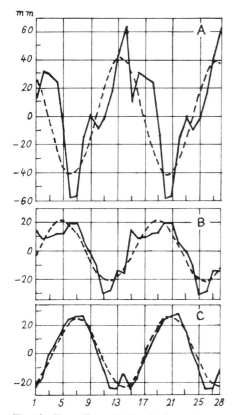

Fig. 13. The effect of Chandler's motion upon the sea level. A. Esbjerg; B. Buenos Aires; C. San Francisco. Solid lines represent the computed Chandler effect and dashed lines the levelled harmonic curves (Maximov, 1970).

Oceans and, in addition, they are opposite in the Atlantic Ocean on the one hand, and the Pacific Ocean on the other hand. The motion of the pole tide from west to east follows the radius vector of the instantaneous axis of the Earth with a retardation of approximately 1.5 months. The magnitude of the retardation was computed by Maximov and Karklin (1965) on the basis of a study on the Chandler effect in the Baltic Sea. The nodal line, which according to the equilibrium theory should be situated along the equator, runs north of it in the Atlantic Ocean, but south of it in the Pacific and Indian Oceans. Maximov has also been able to establish that the maximum amplitude of the pole tide occurs in the middle latitudes of the two hemispheres. This amplitude is, however, somewhat higher than predicted by the equilibrium theory, i.e., at about 60°N and 60°S. At these latitudes the amplitude reaches a magnitude of 4 cm. Comparing this value with the amplitude based on the equilibrium theory, it may be noted that the ratio between the observed and the theoretical amplitudes is approximately 5. In the Baltic Sea this ratio is still more marked, exceeding 7. A chart representing the distribution of the amplitudes and phases (in months) of the 14-month cycle in the world oceans, based on the results of the analysis of sea-level data for 214 stations, is given in Fig. 14 (Maximov, 1970). In this connection it must especially be pointed out that, considering the above results, the Chandler effect must always be kept in mind when seasonal variations in sea level are studied on the basis of records for a single year or even for a few years. Only by taking into account the disturbing effect of the pole tide upon the seasonal cycle may this effect be eliminated.

There seems, however, to be considerable deviation in the results achieved by different oceanographers, probably depending principally on the selection of the stations and periods. Haubrich and Munk (1959) showed that the average amplitudes of the pole tide for the 11 sea-level stations investigated by them were about twice as large as the corresponding amplitudes determined on the basis of the equilibrium theory. Since, as shown by Baussan, the amplitudes of the Chandler effect vary considerably with time, it is rather difficult to draw definite conclusions in this direction. Also, according to the results of Shimizu (1963), based on series of sea-level data referring to the Japanese coasts and covering periods of up to 60 years, the differences in amplitude may in some cases reach 30 mm.

Jessen (1964) chose a quite different method, that of a phase diagram. In this case only the knowledge of the approximate value of the period is required. If the expected period results then the probability is high that the phenomenon is a reality. The results achieved by Jessen are given in Table X. The average period of 434.6 days corresponds to 14 months of 31 days. This period is somewhat longer than the period considered by other scientists, but close enough to prove the reality of the phenomenon.

The Chandler effect seems thus to be an interesting problem which is worth studying in more detail, in particular for the Mediterranean-type seas, where it appears to be still more pronounced than in the oceans. It must also be taken into account that fairly large amplitudes have been mentioned in some of the relevant papers. These results show that the pole tide is a phenomenon which may not be neglected in the studies connected with

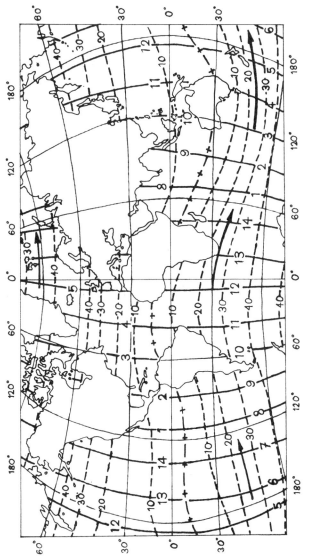

Fig. 14. The distribution of the 'pole tide' in the world oceans. Solid lines = the approximate values of the co-tidal months. Dashed lines = the amplitudes in mm (Maximov, 1970).

TABLE X

THE CHANDLER PERIOD (IN DAYS) FOR DIFFERENT STATIONS, ACCORDING TO JESSEN (1964)

Station	Chandler period (days)
Helsinki	434.6 ± 1.1
Esbjerg	433.0 ± 1.3
Den Helder	434.7 ± 0.6
Marseilles	434.0 ± 1.9
Bombay	438.3 ± 1.1
Sydney	435.4 ± 2.1
Honolulu	433.1 ± 0.6
San Francisco	432.6 ± 1.4
Charleston	435.6 ± 2.9
Average	434.6 ± 1.4

a number of problems, such as the general water circulation in the oceans and the determination of mean sea level.

There may be a certain relationship between the Chandler effect and the variations of the speed of the Earth's rotation. These variations are extremely slight, but they are firmly established. According to Munk and Revelle (1952) the changes in the rotational speed may be explained as a consequence of the secular changes of the atmospheric circulation and especially of those of mean sea level in the oceans. Maximov (1970) advanced the following hypothesis for the variation of the speed of the Earth's rotation. Secular changes in solar activity bring about secular fluctuations in mean sea level. These fluctuations, which may deviate in the different parts of the world oceans, cause changes in the rotational regime of the Earth. These changes are accompanied by a slight change in the shape of the Earth. The changes in the shape of the Earth are the causes of the occurrence of long-period and secular variations of the instantaneous axis of the Earth's rotation and consequently also of the Chandler effect. This hypothesis requires additional studies before it may be accepted without restriction. However, the problem of the relationship between the changes in the rotational system of the Earth and the variations in mean sea level may be briefly described here.

Maximov and Smirnov (1964) have, on the basis of data given by Stovas (1951), and reproducing the changes of the speed of the Earth's rotation and consequently the changes of the length of the day, computed the corresponding fluctuations in mean sea level of the world oceans. Starting from the latitudinal part of the potential of the centrifugal force of the Earth, expressed by the equation:

$$U = \omega^2 a^2 \frac{1}{6} (1 - 3 \sin^2 \varphi)$$

Maximov determined the changes of the potential as the consequences of the variation of the rotational speed. These changes of the potential are:

$$\Delta U = \omega a^2 \frac{1}{3} (1 - 3 \sin^2 \varphi) \Delta \omega$$

where $\Delta\omega = -\omega(\Delta T)/T$. The symbols in the above equations have the same significance as in the equations for the Chandler effect (p.52), while T is the length of the days in seconds and φ the latitude.

According to Stovas the average length of the day has varied during the three last centuries within the limits of -0.005479 and $+0.004658$ sec. Using the formula:

$$\Delta H = \frac{U}{g} \, (1 + k - h)$$

Maximov has been able to compute that the maximum changes in potential cause, for the latitude of $90°$ where the variations in sea level are at their largest, deviations in mean sea level between -5.4 and 4.6 mm. For these determinations Maximov assumed that in the oceans the rates of the variations in mean sea level are 8.7 times larger than those corresponding to the equilibrium theory. The data may therefore be considered as an absolute maximum of the variations. The anomalies of the mean sea level, which are the consequence of the uneven speed of the Earth's rotation, are thus fairly weak, not exceeding 1 cm. It is not possible to prove the occurrence of such slight variations at some locality in the world oceans, since changes in sea level caused by other factors are considerably more pronounced. However, it must be taken into account that the 'rotational' changes in mean sea level cover the whole Earth and are acting during prolonged periods. It is therefore highly probable that the changes in the speed of the Earth's rotation influence the general circulation in the oceans. The problem can hardly be solved at the present time by theoretical speculations. Empirical investigations are doubtless the best approach to this far-reaching question.

CHAPTER 3

THE METEOROLOGICAL AND OCEANOGRAPHIC CONTRIBUTION TO SEA LEVELS

ATMOSPHERIC PRESSURE AND SEA LEVEL

Changes in atmospheric pressure influence the sea level which, at least theoretically, reacts like a reverse barometer. In these theoretical cases a sufficiently long time-span is required in order to allow the sea level to adjust to the effect of the atmospheric pressure. In a stationary case the sea surface should be depressed by 1 cm for every rise in atmospheric pressure of 1 mbar, or to be more exact of 1.005 mbar, and vice versa. In reality, however, this is not the case, since variation in atmospheric pressure is associated with change in the direction and velocity of the wind, which is one of the most important factors contributing to the fluctuations in sea level.

The distribution of air masses over the oceans is not constant; it changes continuously and is, in addition, characterized by a marked seasonal variation. A minor part of this variation is the consequence of the seasonal changes of the total atmospheric pressure over the oceans and seas from 1,012 mbar in December to 1,014 mbar in July, due principally to the shift of air masses towards Siberia in winter. This change cannot contribute materially to the variations in sea level and must therefore be eliminated from the data when examining sea-level observations covering a time-span of one year or more. Pattulo et al. (1955) were probably the first oceanographers to carry out such an elimination.

Assuming that P_n is the average monthly atmospheric pressure at a given locality in an oceanic area and P_o the average atmospheric pressure over all oceans for the same month, then $c' = P_n - P_o$ corresponds to the theoretical responce of the water surface relative to the mean sea level of the total area covered by sea water. The sea-level records must thus be corrected by this amount in order to eliminate the effect of atmospheric pressure. The base level being arbitrary, it is more adequate to use the correction factor $c = c' - \overline{c}$, which corresponds to the mean annual value of zero. The scheme giving the relevant computations is illustrated in Table XI. This procedure should be applied in all cases where data are based on monthly means and where more accurate results are required.

The monthly values $P_o - 1000$ mbar, designating the average pressure over all oceans, are of course identical for all localities. They were determined by Pattulo et al. (1955) on the basis of values for each $5°$ quadrangle between the latitudes of $90°N$ and $10°N$ and for each $10°$ quadrangle between the latitudes of $10°N$ and $60°S$.

Although the pronounced contribution of atmospheric pressure to the seasonal cycle in sea level was fully recognized in the 1950's, the mechanism of the effect had not been

TABLE XI

THE SCHEME FOR ATMOSPHERIC PRESSURE CORRECTION

	J	F	M	A	M	J	J	A	S	O	N	D
(P_n-1000)mbar	6.5	8.0	9.0	9.5	14.5	14.0	12.5	10.0	10.5	8.5	7.0	6.0
(P_0-1000)mbar	12.4	12.4	12.8	12.8	12.8	13.5	14.0	13.3	13.4	12.8	12.6	11.9
(P_n-P_0)mbar	−5.9	−4.4	−3.8	−3.3	1.7	0.5	−1.5	−3.3	−2.9	−4.3	−5.6	−5.9
c cm	−2.7	−1.2	−0.6	−0.1	4.9	3.7	1.7	−0.1	0.3	−1.1	−2.4	−2.7

studied at that time on a world-wide scale. A number of additional studies were therefore necessary. The fact that the comprehensive Marine Atlas (*Morskoi Atlas*, 1953) had become available to a large circle of scientists, made it possible also to compute more accurate data for the average monthly values of atmospheric pressure over the oceans. The Marine Atlas gives charts illustrating the distribution of the monthly values of atmospheric pressure over the oceans for the latitudes situated between $70°N$ and $60°S$. In addition, for the months January and July this distribution is given for the total surface of the Earth's globe lying between $84°N$ and $78°S$. On the basis of these two charts the atmospheric pressure was computed for points corresponding to every 10 latitudinal and longitudinal degrees. Since the charts did not extend to the poles, an extrapolational process has been necessary. The procedure as a whole was a little inaccurate but the final result was satisfactory (Lisitzin, 1960). For January the global mean value of atmospheric pressure was determined to be 1,011.47 mbar and for July 1,011.27 mbar. The departure was thus only 0.2‰.

Considering next the conditions characteristic of the sea-water covered areas a considerable difference could be noted immediately. The average atmospheric pressure was in this case 1,010.33 mbar in January and 1,012.22 mbar in July. The deviation thus reached approximately 2‰ of the total value, being of the same magnitude as that given by Patullo et al. The montly average values of atmospheric pressure $(P - 1000$ mbar) computed on the basis of the charts given in the Marine Atlas are the following:

J	F	M	A	M	J	J	A	S	O
10.33	10.33	10.64	11.39	11.18	11.49	12.22	11.69	11.54	11.18

N	D	Year
10.55	10.22	11.06

Since the investigation is based on monthly sea-level means, it may be assumed that the surfaces of the oceans and seas react like a reverse barometer, since the sea level has sufficient time to adjust itself to the outer effect. The increase in atmospheric pressure corresponding to 1 mbar should thus result in a decrease of sea level of 1 cm, and vice versa.

It is a well-known fact that the average atmospheric pressure over the oceans — and self-evidently also over the continents — differs considerably in individual regions. The natural consequence is that the mean sea level shows a corresponding, but reverse, configuration. The height differences between regions with extreme values of mean sea

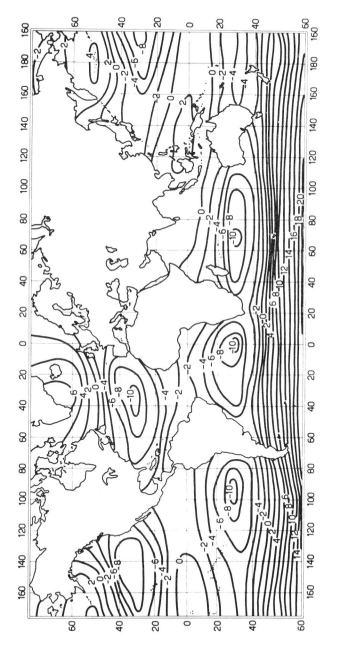

Fig. 15. The distribution of the average sea-level differences (in cm) caused by atmospheric pressure (Lisitzin, 1961b).

level amounts to 32 cm. The regional distribution of the average annual sea-level differences, caused by the effect of atmospheric pressure, is reproduced in Fig. 15 (Lisitzin, 1961b). The chart in this figure shows that the average maximum sea level occurs in the southern hemisphere around the Antarctic and that it is extremely pronouced. The Icelandic and Aleutian low-pressure areas are responsible for the corresponding regions of high sea level. Fairly marked sea-level minima occur in five different parts of the worlds oceans. They are in each case the consequence of an atmospheric pressure high, being situated around the Azores, in the South Atlantic, the North and South Pacific and the South Indian Oceans. In spite of the fact that the atmospheric pressure highs fluctuate in strength and position in different months, they prevail troughout the whole year.

It is of considerable interest to note, that the seasonal fluctuations in atmospheric pressure bring about a shift of the water masses between the hemispheres and between different latitudinal zones. For instance, the quantity of water which passes the equator from the middle of October to the middle of January may be estimated at 3,150 km^3 (Lisitzin, 1960). This water quantity, although insignificant in comparison with the total water masses in the oceans, may have a considerable effect on sea-level fluctuations.

In addition to the average yearly distribution of sea-level heights in the world oceans as the consequence of the distribution of atmospheric pressure, the problem of the motion of the water masses in response to the changing effect of this pressure in the course of the year may be of considerable interest. In this connection all other contributing factors have been left out of the following account. In order to get a perspicuous picture, the time of the occurrence of the minimum of atmospheric pressure, corresponding to the maximum sea level, has been computed by harmonic analysis for a grid of points situated at a reciprocal distance of 10°.

The results of the harmonic analysis are reproduced graphically in the chart in Fig. 16. The solid curves in this chart correspond to the time when the maximum sea level was reached, and they mark the first day of every month. Following the practice in tidal researches, these curves could be referred to as co-tidal lines, although the use of this term is not correct in our case, since the changes in sea level are not related to tidal motion. The dashed curves represent the amplitudes expressed in cm. They may, therefore, be called co-range lines. Since the amplitudes amount to half of the total range of variation, Fig. 16 shows that the most marked seasonal variation in sea level, due to the changes in atmospheric pressure, exceeds 16 cm. On the other hand, it may be noted that over extensive areas this variation is less than 4 cm.

Fig. 16 indicates not less than six amphidromic regions. In the North Atlantic Ocean there is one region of this type in the eastern and another in the western part of the basin. In the following the former will be referred to as the Iberian and the latter as the Cape Hatteras amphidromic region. In the Pacific Ocean proper and the adjacent sea areas the number of amphidromic regions is four: the Californian, Wake Island, Indonesian and New Zealand amphidromic regions. In the Indian Ocean no amphidromic areas have been found.

It is not always very easy to determine the exact positions of the amphidromic points..

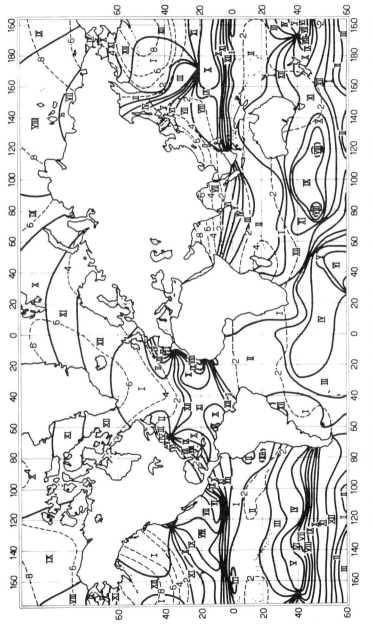

Fig. 16. The times for the arrival of the highest sea level due to the annual cycle of atmospheric pressure, given by solid curves, and the amplitudes (in cm) of the seasonal variation in sea level due to the same effect, given by dashed curves (Lisitzin, 1961b).

TABLE XII

THE AMPLITUDES (IN CM) OF THE VARIATION IN SEA LEVEL IN THE IMMEDIATE VICINITY
OF THE AMPHIDROMIC POINTS AND IN DIFFERENT DIRECTIONS FROM THEM

Amphidromic region	Position of minimum amplitude		Minimum	Directions from the amphidromic points							
				N	NE	E	SE	S	SW	W	NW
Iberian	40°N	10°W	0.7	3.3	–	–	–	–	1.3	1.8	4.7
Cape Hatteras	40°N	70°W	0.5	–	–	1.7	0.6	0.8	–	–	–
Californian	40°N	130°W	0.7	–	–	–	2.6	1.3	1.6	4.2	5.1
Wake Island	20°N	170°E	0.3	1.1	2.0	0.7	0.7	0.8	1.0	1.4	0.7
Indonesian	0°	120°E	0.2	3.0	2.3	1.7	0.9	2.2	2.1	2.2	0.6
New Zealand	40°S	170°W	0.3	0.8	0.6	0.5	2.1	1.9	1.3	1.3	0.5

Nevertheless, the rapidly decreasing amplitudes indicate the approximate position with an accuracy of within a few degrees. Table XII indicates that for all amphidromic points the computed amplitude is less than 1 cm, and practically everywhere about half of the least marked amplitude in the neighbourhood of this point. The only exception is the amphidromic point off Cape Hatteras which is rather weakly developed and the determination of its exact position has therefore been more difficult than in the other cases.

The question of the causes which give rise to the amphidromic regions is principally a meteorological problem, but it cannot be left completely out of consideration here. It seems to be reasonable to start from the assumption that the amphidromic points originate in areas where the atmospheric pressure varies during the particular seasons, but where the core remains practically unaffected by these changes. For instance, in the region around the Azores the conditions seem to be fairly favourable in this respect. An atmospheric high prevails there during the whole year, although the position of its centre varies markedly from one month to another. Generally the high-pressure centre is situated further away from the continental coast during the warm season than in winter. The migration towards the west is accompanied by an increase in the strength of the high, resulting in the fact that between the anticyclone and the coast there always remains a spot where the atmospheric pressure is more or less unaltered. It is around this spot that the Iberian amphidromic region is generated.

A practically analogous explanation to that outlined above may be given concerning the Californian amphidromic region. In this case also the origin is the consequence of the migration of the centre of the atmospheric high and of its seasonal variation in the strength of this high occurring in the North Pacific Ocean.

Off Cape Hatteras the average monthly atmospheric pressure is relatively consistent during the entire year, but the marked changes in the direction of the wind from season to season indicate that an annual difference occurs in the distribution of atmospheric pressure over a larger area. In fact, there is a secondary pressure high north of Florida in January, which cannot be traced in July.

The Wake Island amphidromic region is also situated in an area where the seasonal changes in the average value of atmospheric pressure are rather insignificant. However, during the winter months it is situated in the transition area between the Siberian and the North Pacific Ocean highs. The former of these highs disappears in summer, while the North Pacific high continuously influences the situation in the area east of Wake Island. The decrease in atmospheric pressure towards the west is now probably the cause of a marked change in atmospheric circulation.

The Indonesian amphidromic region around Borneo is probably the most interesting case, with a stable core but considerable seasonal variations in atmospheric pressure. During the winter half of the year the entire area is dominated by the effect of the pronounced Siberian high and the Australian low, the prevailing wind directions during this season being northeast and northwest. Conversely, in the summer months atmospheric pressure increases towards the south within this region, resulting in the predominating wind directions from the southwest and southeast.

The amphidromic region situated to the east of New Zealand seems to be brought about by less distinct variations in atmospheric pressure. Nevertheless, in this case also a slight change in the average atmospheric pressure of the core and a considerable shift of the direction of the wind in the area may be noted from winter to summer.

The chart in Fig. 16 shows very distinctly the differences in sea level between the winter and the summer seasons. Thus the sea level is generally high in the central parts of the oceans in the winter of the northern hemisphere. This fact is highly accentuated in the Atlantic Ocean, where the maximum occurs in the middle parts of the ocean and occasionally along the coasts between the latitudes of 30°N and 60°N in January, while in the southern hemisphere there is an extensive region, covering practically the entire subtropical zone, characterized by a maximum sea level in February. In the northern parts of the Pacific Ocean the two winter months December and January are predominant. In the central regions of this Ocean, the sea level reaches its maximum in December, but there is a retardation of one month in the areas situated farther east and west. In this ocean also the northern and southern borders of this region run approximately along the latitudes of 60°N and 30°N. In close conformity with the conditions in the Atlantic Ocean, the maximum sea level is noted in the southern parts of the Pacific Ocean in February. In this ocean the concerned region is, however, somewhat narrower, extending only from the equator to the latitude of 20°S. Farther south the picture is affected by the water circulation within the New Zealand amphidromic region which perturbates the distribution and so involves additional features. Owing to its restricted extension to the north of the equator, the Indian Ocean deviates from the two oceans described above. South of the equator, the region corresponding to the maximum sea level in February extends further southwards than in the Pacific Ocean, but not as far as in the Atlantic Ocean. The regions to the north of the equator in the Indian Ocean are in keeping with the coastal areas in the northern parts of the other oceans. In the Arabian Sea and the Gulf of Bengal the highest sea level is reached in July, which, in addition to June, is the most probable month for the occurrence of the maximum height in sea level along the coasts of the northern parts of the Atlantic and Pacific Oceans.

Along the equator or more precisely, within the zone extending from the equator to the latitude of approximately 10°N, the 'co-tidal' lines in the Pacific Ocean run very close to one another. In the middle part of the zone the highest sea level is reached in autumn. Another case, which is not, however, as pronounced as the former of more-or-less parallel 'co-tidal' lines occurs in the South Pacific Ocean to the east of the New Zealand amphidromic region, extending between the latitudes of 40°S and 50°S.

The northernmost and the southernmost regions in Fig. 16 are also of considerable interest. In the Arctic Ocean (Lisitzin, 1961a) the highest sea level, due to the atmospheric pressure, is noted off the Siberian coast in August, in the central parts of the ocean in September, to the east of Greenland in October and, finally, off the Norwegian coast in November. Around the Antarctic continent the general features seem to be reversed. In this marine region the maximum sea level is reached in May and June, approximately at the same meridian as at which it occurs in the Arctic Ocean in November, corresponding thus to the difference in boreal and austral seasons. However, the data are, for the most part, too indefinite to allow the exact allocation of regions most characteristic of the sea-level maximum heights in the Antarctic area in winter. On the whole, a coincidence of longitude with the summer maximum in the Arctic zone is by no means excluded. There is another feature in the southern subpolar region which may be pointed out. The time of the occurrence of the highest sea level does not indicate pronounced regularities as in the other regions. Of course this feature may be due to the fact that the data on which the chart representing atmospheric pressure is based are much scantier for the circumpolar zone around the Antarctic than for the other parts of the oceans. On the other hand, the range of the variations is fairly weak and the significance of the fluctuations therefore only slight.

A closer examination of the co-range lines given in the chart in Fig. 16 reveals a number of interesting features. A certain relationship may, for instance, be discerned between the amplitudes of the variation and the extension of the area characterized by the occurrence of the maximum sea level during a definite month. This relationship is, of course, by no means surprising. If the high- or low-pressure areas are large, they are simultaneously fairly pronounced, and this fact must be reflected in the amplitudes of the sea-level fluctuations. In each of the three extensive oceans there is a region where the amplitudes are marked. In the North Atlantic Ocean this region encircles Iceland; in the North Pacific Ocean its core is situated around the Aleutian chain; and in the Indian Ocean the largest amplitudes are noted in the Arabian Sea and the Gulf of Bengal. All these areas are affected by deep-pressure lows: the Icelandic and the Aleutian cyclones in winter and the South Asiatic cyclone in summer. The considerable amplitudes of the seasonal sea-level oscillations in the Artic Ocean off the coast of eastern Siberia are, on the contrary, the consequence of the strong North Asiatic high-pressure zone in winter or, to be more precise, of its far-reaching effects.

To the south of the equator the seasonal variation in atmospheric pressure is, as a rule, less pronounced than in the northern hemisphere. The largest fluctuations in atmospheric pressure occur over the continents and they are therefore of minor concern in this connection. The principal characteristic of the regional distribution of atmospheric pressure over the oceans is a belt of high pressure, the most pronouced part of which is situated between the latitudes 20°S and 30°S. During the summer half-year this belt is much wider than during the winter half-year and the extension of the zone is accompanied by a pronounced increase in atmospheric pressure. This feature is self-evidently distinctly reflected in the regions where sea-level variations have an amplitude exceeding 2 cm. In the Atlantic and the Indian Oceans these zones extend between the continents from one coast to another, while in the Pacific Ocean the belt is interrupted in the central parts of the basin. This phenomenon cannot be ascribed exclusively to the fact that the Pacific is considerably wider than the other oceans. The position of the core of the high-pressure area in the South Pacific Ocean, being situated not in the central parts of the basin but farther to the east off the coast of South America, must also be taken into consideration. In addition, it must be borne in mind that in the South Pacific Ocean the differences in atmospheric pressure between winter and summer within the high-pressure belt are relatively slight in comparison with those recorded within the corresponding zones in the South Atlantic and South Indian Oceans.

The above-described results are in many respects interesting and significant, but they are, of course, purely theoretical, since the effect of atmospheric pressure is only one of the numerous elements which may contribute to the fluctuations in sea level. Generally it is not possible to determine that part of the variations associated with one definite factor, since the different terms affect each other. Not until the local and regional influence of other contributing factors, especially that of wind stress, is known in more detail, will it be possible to arrive at a fuller understanding of the characteristics of the recorded sea-level data. The contribution of the seasonal changes in the specific volume of sea water to the variation in sea level will be discussed later (pp. 86–90).

The response of sea level to the changes in atmospheric pressure has been investigated in numerous cases in order to obtain the numerical ratio of this relationship and its deviations from the theoretical value. As already mentioned, this value is practically equal to unity, if the atmospheric pressure is expressed in mbar and the sea-level variations in cm. In older papers and meteorological yearbooks the atmospheric pressure was expressed in mm, with the consequence that the theoretical ratio was −13.2, if the sea level was given in cm. Different authors have treated the problem in different ways. In a number of cases the results have been based on daily observations, but in others monthly averages have been utilized. The two series of corresponding data have either been compared with each other as such, or the consecutive differences of the connected values have been used as the basis for the investigations. Rouch (1944), for instance, chose the latter method. Starting from the differences in the observations from one day to the next, Rouch produced, for the data collected during the cruise of 'Pourquoi Pas?' in the Antarctic region, the following results: the correlation factor between the atmospheric pressure (in mm), on the one hand, and the sea level, on the other hand, was 0.94, while the concerned ratio amounted to −13.9.

Following the method outlined by Rouch, the ratio for the relationship between the

monthly differences for the two elements has been computed for some stations in the
Mediterranean (Lisitzin, 1954b). The results were surprisingly good, i.e., −13.0 for
Monaco and −13.5 for Porto Maurizio. However, it must be pointed out that the ratio
varied considerably for the particular cases and that only for prolonged series − in the
cases mentioned above the observations covered 10 years − did the average ratio
approach the theoretical value. In such cases the assumption may be made that the
disturbing effects of other contributing elements have been more or less eliminated from
the data.

Although the examples given above have indicated that the method introduced by
Rouch has achieved satisfactory results, the direct comparison of data on atmospheric
pressure and sea level or, to be more exact, of the departures of these data from the
average values, is the more commonly applied procedure. The studies and their results
may in this case be divided into two groups, mainly referring to the period on which the
data have been based. If the data have been based on daily observations the ratio
frequently shows the tendency to reach values which are considerably higher than the
theoretical one. However, it must be emphasized that in most of these cases the
deviations in atmospheric pressure have been pronounced, with the consequence that the
contribution of wind stress, generally acting in the same direction as the atmospheric
pressure, was also very pronounced, thus highly distorting the relationship.

More numerous are, however, the studies on the relationship between the atmospheric
pressure and sea level which have been referred to monthly average data. In these
particular cases the ratio has been almost invariably lower than the theoretical unity. It
may be appropriate to mention some of these studies and their results.

Galerkin (1960) investigated the part corresponding to the changes in atmospheric
pressure in mbar in the total sea-level variations in the Sea of Japan. His results were, that
the ratio between the two factors varies from −0.37 to −0.68 for the particular stations
being, on average, −0.44. On the other hand, the variability of the ratio is considerable for
the separate months and no general rule could be deduced.

Lisitzin (1961a) studied the ratio between the amplitudes in sea level, which could be
ascribed to atmospheric pressure (in mbar), and those recorded by gauges in the Arctic
Ocean. Although this sea basin is one of the regions where the effect of water density is
relatively weak and the contribution of atmospheric pressure fairly marked, the concerned ratio was, on an average, −0.30 for the 9 stations situated in the area. It may therefore
be mentioned that additional studies (Lisitzin, 1964c) have shown that there is a pronounced seasonal variation in the inflow of water into the Arctic Ocean from the Atlantic
Ocean, and to some less degree also from the Pacific Ocean, and that this inflow is of
decisive significance for the annual sea-level fluctuations in the basin. The complexity of
the phenomenon must therefore be emphasized once more.

Moreover, Pattullo et al. (1955) considered the contribution of atmospheric pressure
to the seasonal variation in sea level. The authors did not make any attempt to determine
the ratio, but paid attention to the magnitude of the effect of atmospheric pressure upon
the changes in sea level. According to the results of Pattullo et al., the effect exerted by

atmospheric pressure is small in comparison with the recorded departures in sea level. In most cases the correction for atmospheric pressure results in a weak reduction in range, the average value of this reduction being 1.6 cm for all examined stations and station groups. Since the material on which the authors based their computations is extremely extensive and cover practically every part of the world oceans with highly varying, but in some cases fairly pronounced, amplitudes in sea-level changes, this result is not very instructive as regards the validity of the theoretical ratio in different regions.

Finally, it may be pointed out that not only different contributing factors are the cause of the deviations between theory and observations. It is inadvisable to take into account only local meteorological conditions. The situation prevailing over more extensive areas and all the relevant variations in time and space must be considered, if accurate and reliable results are required.

THE WIND EFFECT – STORM SURGES

The effects of the wind upon sea level may be of many different kinds. To begin with, it may be mentioned that wind produces waves, which self-evidently influence the sea level. However, the short-period variations in sea level which are the consequence of ordinary wave motion do not generally belong to the field of sea-level research, and they are therefore left out of consideration here. In addition, mention must be made of the wind-driven currents, a phenomenon to which attention will be paid in the section dealing with the effect of currents upon the sea level (p.90ff). Finally, there is the direct effect of wind stress upon the sea surface. This effect is the cause of a great number of departures of sea level from the height of the predicted tide, or, in basins where there is no tide worth mentioning, from the average sea level. These departures are brought about partly by transient winds which are associated with rapidly moving storm formations and partly by more steady winds. In addition to the adjustment to the varying atmospheric pressure, the fluctuations in sea level may in these cases be attributed to the accumulation or depletion of water along the coast owing to the effect of the tangential stress of the wind force upon the water surface.

The effect of the tangential stress of the wind upon the sea surface brings about a raising or a lowering of the sea level. In order to describe this phenomenon for the more extreme cases the term 'storm surge' is generally used. According, for instance, to Heaps (1967) the raising of the water level, is denoted as a positive surge and the corresponding lowering as a negative surge. Instead of the term 'storm surge' the term 'set-up' of water has been recommended in some cases (Miller, 1957). In this case one may also speak of positive and negative set-ups. In some less pronounced cases the term 'piling-up' describes the phenomenon, usually as a substitute for the commonly used German term 'Windstau'. Conversely, the term 'wind tide' is hardly adequate here, since it could imply the possibility of a perodic phenomenon.

The average duration of a storm surge, as a rule consisting – when positive – of the increase in sea level to a peak which is followed by a decrease in level, covers a time-span extending from a few hours to two or three days. Heaps mentions, for instance, that a number of storm surges in the North Sea are diurnal in character, while those occurring on the west coast of Great Britain are generally of shorter duration, covering a period of 9–15 hours. Since the meteorological contribution to the sea level is superposed upon the normal astronomical tide, the determination of the magnitude of the storm surge at a given locality from the recorded sea-level heights consists in the elimination of the tidal effect. This procedure may be done in different ways and one of these methods, for instance, has been described by Rossiter (1959). The substraction of the predicted tidal height from the recorded sea-level readings at the same locality and time yields a value which may be described as the residual. All the residuals corresponding to the time of occurrence of the storm surge at a given coastal station, let us say, for every hour, reproduce the heights and the general character of sea-level departures caused by the storm surge. This method has so far only limited application, since it depends decisively upon the reliability of the predicted tide. In the cases where the tide is complex in character and, especially if the perturbating effect of shallow water is pronounced, the residuals will not give an adequate picture of the different phases of the storm surge.

A storm surge of considerable proportions is generally the consequence of the passage of a deep atmospheric depression across the part of the sea basin in which the surge arises. As soon as the depression approaches the locality under consideration, the atmospheric pressure decreases and the sea level rises. When the depression leaves the area, atmospheric pressure increases again and the sea level decreases. If the tropical cyclones, which are very intense, are left out of consideration, the contribution of the effect of atmospheric pressure to the height of the storm surge at a given locality is, as a rule, rather slow. The more accentuated variations in sea level may therefore be mainly attributed to the winds associated with the depression. The tractive force of the wind upon the water surface results in the water being dragged in the direction of the wind. Owing to the influence of the Earth's rotation, the moving water masses are deflected to the right in the northern hemisphere and to the left in the southern hemisphere (p.92). When these water masses reach the coastline, the sea level increases rapidly, since the water is piled up, resulting thus in a positive storm surge or set-up. If the net water transport is directed away from the coast, the result is the lowering of the sea level, which is described as a negative storm surge or set-up.

The most pronounced storm surges generally arise in relatively shallow sea regions bordering the oceans and in shallow marginal and Mediterranean-type seas. The most severe storm surges which may occur along the oceanic coasts are the hurricane surges which ravage the coasts of the Atlantic Ocean and of the Gulf of Mexico in the United States, and typhoon surges which have a highly devastating effect along the Pacific Ocean coast of Japan. The west coast of the British Isles also experiences the destructive effect of open-coast storm surges from time to time. In the shallow Bay of Bengal storm surges may also sometimes develop which result in devastating floods accompanied by loss of life and property. Among the storm surges which strike the coasts of more or less enclosed sea regions, those occurring within the North Sea and the Baltic Sea basins are probably the most investigated and best known.

. The slope of the water surface which is the consequence of the stress of a steady wind acting over the sea surface has been determined theoretically by different authors, among whom special reference must be made at least to Nomitsu (1935) and Hellström (1941). Later studies based on a considerable amount of observed or recorded sea-level data have shown that the formula, which has been deduced theoretically, corresponds relatively closely to actual conditions.

The slope of the water surface $\partial \xi / \partial x$ due to the wind stress τ is determined from the formula:

$$\frac{\partial \xi}{\partial x} = \frac{\lambda \tau}{g \rho_w h} \tag{1}$$

where ξ is the elevation of the water surface and x the horizontal co-ordinate in the direction of the wind. g is, as usual, the acceleration of gravity, ρ_w the density of the water and h the average depth of the sea basin. The factor λ varies between the limits 1 and 3/2, depending on the depth and the frictional conditions at the sea bottom. If the sea basin is not too shallow, it may be assumed that λ is close to unity.

Considering next the relationship between the wind stress τ and wind velocity W at the 'anemometer level', corresponding approximately to 10 m, no complete agreement has so far been reached between the results achieved by different authors. Nevertheless, if the wind velocity is larger than 6–8 m/sec, the quadratic relation:

$$\tau = k \rho_a W^2$$

seems to be the most probable (Munk, 1947; Sverdrup, 1957). In this equation k is a drag coefficient depending upon the roughness of the water surface and ρ_a the density of the air. The possible dependence of k on the wind was investigated by Neumann (1948). However, so far no generally accepted conclusions are available. For the Gulf of Bothnia in the northern part of the Baltic, in cases of strong winds, the drag coefficient was determined by Palmén and Laurila (1938) as $2.4 \cdot 10^{-3}$. A similar value was computed earlier by Ekman (1906) using Colding's results for the piling-up of the water during a strong storm in the southern Baltic.

Assuming that the average density of the air is $1.3 \cdot 10^{-3}$ g/cm^3 and that of the water (of the Baltic) 1 g/cm^3, and using for g the value 980 cm/sec^2 and for λ the value 1, eq. 1 may be written:

$$\frac{\partial \xi}{\partial x} = \frac{3.2 \cdot 10^{-9} W^2}{h}$$

where the velocity of the wind is expressed in cm/sec. If the increase in sea level (cm) for the distance of 100 km in the direction of the wind is denoted by ΔH and if the wind is given in m/sec, while the depth h is expressed in m, the following practical formula representing the slope of the water surface caused by the piling-up effect of the wind is obtained:

$$\Delta H = \frac{\alpha W^2}{h} \tag{2}$$

where α has the value 3.2. It must be taken into consideration that the value of this coefficient depends to some extent upon the density of the air, which, being a function of temperature and atmospheric pressure, may vary within the limits of approximately 10%.

The wind-stress-produced slope of the water surface represented by eq. 2 implies that the total wind stress at the sea surface and the bottom stress are balanced by the pressure gradient in the water column. In the cases where the sea depth is considerable, the bottom stress may be neglected. The wind stress determines a flux of momentum from the air to the water. If there is a steady state situation, the flux of momentum must be equal to the flux from the limited sea to the boundaries resulting from the pressure difference Δp due to the height difference ΔH, the pressure difference being:

$$\Delta p = g \rho_w \Delta H$$

The validity of the formula (2) has been proved in many cases in the Baltic proper and in its gulfs. The formula will later be used (pp.80–86) for the determination of the damping effect of the continuous ice cover upon the wind-stress-produced pilling-up of the water surface in the northern parts of the Baltic Sea area.

For more complicated cases of storm surges, simplified formulae are hardly more valid and the computations must be based on hydrodynamic differential equations which represent the motion of the sea as functions of the atmospheric pressure and wind stress. The theoretical studies of storm surges are based on the equation of continuity:

$$\frac{\partial \xi}{\partial t} + \frac{\partial U}{\partial x} + \frac{\partial V}{\partial y} = 0 \tag{3}$$

and the equations of motion:

$$\frac{\partial u}{\partial t} - fv = -g \frac{\partial \xi}{\partial x} - \frac{1}{\rho_w} \frac{\partial p_a}{\partial x} - \frac{1}{\rho_w} \frac{\partial F}{\partial z} \tag{4}$$

$$\frac{\partial v}{\partial t} + fu = -g \frac{\partial \xi}{\partial y} - \frac{1}{\rho_w} \frac{\partial p_a}{\partial y} - \frac{1}{\rho_w} \frac{\partial G}{\partial z} \tag{5}$$

where:

$$U = \int_0^h u \, dz , \qquad V = \int_0^h v \, dz$$

and:

$$F = -\mu \frac{\partial u}{\partial z} , \qquad G = -\mu \frac{\partial v}{\partial z}$$

In these equations given by Proudman (1954b), x, y and z represent the Cartesian co-ordinates with x and y, as usual, horizontal and z vertically downwards. $z = 0$ corresponds to the undisturbed water surface and $z = h$ to the sea bottom. t refers to the time. u and v are the components of the water motion at depth z, in the directions of increasing x and y respectively. ξ is, as above, the height of the sea level above the undisturbed water surface. p_a gives the atmospheric pressure. F and G are the components of the frictional stress which the water masses above the depth z exert on the water column below this depth, in the directions of increasing x and y respectively. The four additional symbols in eq. 3–5 are generally assumed to be constant for theoretical purposes. Of these symbols μ represents the coefficient of eddy viscosity, ρ_w the density of the water, f is the geostrophic coefficient ($f = 2\omega \sin\varphi$, where ω is the angular speed of the Earth's rotation and φ the latitude) and g the acceleration of the Earth's gravity.

The frictional stress components at the surface F_s and G_s in the directions x and y are determined by the surface boundary conditions:

$$-\mu \frac{\partial u}{\partial z} = F_s , \qquad\qquad -\mu \frac{\partial v}{\partial z} = G_s , \qquad\qquad \text{at } z = 0$$

At the sea bottom the conditions are correspondingly:

$$-\mu \frac{\partial u}{\partial z} = F_b , \qquad\qquad -\mu \frac{\partial v}{\partial z} = G_b , \qquad\qquad \text{at } z = h$$

where F_b and G_b are the components of the bottom friction given as a function of bottom currents.

The first, classic, study of the interaction between wind stress and atmospheric pressure on the one hand and sea level on the other hand was made by Proudman and Doodson (1924b). The authors left the geostrophic effect out of consideration and determined the variations in sea level in a closed rectangular sea basin. The motion of the water was considered to be two-dimensional in the vertical xz-plane, while the boundary conditions corresponded to a zero horizontal current at the vertical ends of the basin and along the horizontal bottom. In a later investigation Proudman (1929) neglected the effect of friction, and in this way solved the hydrodynamical equations for ideal sea basins, obtaining the currents and sea-level changes produced by the variations of atmospheric pressure. Nomitsu (1934) chose as a starting point a long straight channel of uniform cross-section and computed the changes in sea level caused by wind stress and an atmospheric pressure gradient continuously acting over the entire length of the basin. Nomitsu also took account of the geostrophic effects and included three types of bottom boundary conditions: zero current at the bottom, zero friction at the bottom and bottom friction being directly proportional to the bottom current.

A number of more recent investigations (Bowden, 1956; Weenink, 1958; Weenink and Groen, 1958) have shown that in sea regions dominated by strong tidal currents the components of bottom friction associated with the occurrence of storm surges may be expressed by the equations:

$$F_b = \beta \rho_w U - m F_s , \qquad G_b = \beta \rho_w V - m G_s$$

where $\beta = r/h$. The value of the coefficient r must be determined empirically and it is a function of x and y. Weenink chose a constant value, $r = 0.24$ cm/sec, for the southern parts of the North Sea. The coefficient m in the equations given above is of the magnitude of 0.1 or less.

The above equations represent a summary of the fundamental equations of motion given by Groen and Groves (1962) in a linearized form. The non-linearity of Groen and Groves' equations is partly the consequence of the assumption that the friction at the sea bottom is proportional to the square of the bottom current. In addition, the non-linear terms take into account the interaction between the storm surges and the astronomical tide, an interaction which is especially important in shallow-water areas in the vicinity of the coasts.

Theoretical investigations of storm surges in the North Sea have been made, for instance, by Van Dantzig and Lauwerier (1960a,b).

These authors considered the North Sea as a rectangular region and the computations were carried out for a basin of uniform depth and for a basin corresponding more closely to the actual features of the North Sea. Uniform wind fields and fields with wind-stress components varying linearly were investigated by Van Dantzig and Lauwerier. Starting from the assumption that the sea was at rest at $t = 0$, the sea-level heights were found to be a function of the northerly wind-stress fields.

In addition to the investigation of storm-surge problems in sea basins corresponding approximately to the configuration of a real sea area, like the North Sea, a considerable amount of research work has been carried out to determine the interaction between non-stationary wind-stress fields and the water surface in idealized basins, generally of constant depth. The types of basins most frequently considered in these investigations are the following: a closed rectangular basin and a rectangular gulf of infinite length (Proudman, 1954a); an infinite sea with no boundaries (Crease, 1956; Lauwerier, 1956a); a sea in the shape of a double wedge, representing to some degree the North Sea—English Channel system (Lauwerier, 1956b).

During the 1960's numerical solutions of the hydrodynamic equations for storm surges rapidly increased in number. The application of different numerical methods has made it possible to analyze theoretically the character of storm surges in the particular sea basins and has achieved relatively satisfactory results, in spite of the introduction of considerable simplifications of depth and boundary conditions in the computations. Although the computations on which the numerical work is based are, as a rule, not complicated, the procedures themselves are time-consuming and the final results are generally reached by a number of successive approximations. The development of high-speed electronic computers has facilitated these investigations to a very high degree. Among the particular studies mention may be made in this connection of the researches by Hansen (1956, 1962), Fischer (1959) and Lauwerier and Damsté (1963) for the North Sea; by Svansson (1959) Laska (1968) and Uusitalo (1960, 1972) for the Baltic; by Rybak (1971, 1972) for the

White Sea and by a number of Japanese oceanographers for the bays situated along the Pacific coast of Japan. An interesting summary of the earlier investigations in this field was given by Welander (1961).

In all the above-mentioned cases the hydrodynamic differential equations were substituted by a number of finite-difference equations giving the values of the elevation in sea level ξ and for the terms U and V, by a system of grid points uniformly covering the entire sea region which was to be investigated. The difference equations form a series of recurrence relations, with the help of which the three factors mentioned above may be determined for a particular time $t = (n + 1) T$ from the corresponding value for the time $t = nT$. Since the values are known for $t = 0$, the values for the subsequent time points $t = T, 2T$, etc., may be determined. In this connection special attention must also be paid to the boundary conditions along the coasts of the relevant sea basin and they must be included in the above-outlined step-by-step procedure. For this purpose a sea boundary consisting of line segments connecting consecutive grid points and approximating as closely as possible to the actual coastlines has to be determined. An important stipulation in this respect is that the whole system of difference equations must be stable and that errors are not continuously increasing with time during the process of the consecutive numerical computations. Different methods have been adopted to ensure the stability of the numerical system, using 'smoothing' operators to the values of ξ, U and V following each step T of the iterative procedure. Although in most of the investigations mentioned above some kind of numerical smoothing has been applied, continuous research is necessary in order to give a more exact picture of the conditions for which stability may be fully insured in the particular systems of difference equations.

One of the most interesting studies on the theory of storm surges was presented by Welander (1957) at a relatively early stage of the development of the hydrodynamic computations of this phenomenon. Starting from the eq. 3–5 and assuming that:

$$U = V = 0, \qquad \text{at } z = h,$$

Welander was able to show that in a shallow sea basin as for instance, the North Sea, the total flow of water at any point may be expressed in terms of the local time-development of the wind-stress field and the surface slope. Subsequently Welander was able to deduce an integro-differential equation for the sea-level elevations. This equation could be solved numerically for a given sea basin and wind-stress field on the basis of the step-by-step technique, formulating for each step the coastal boundary conditions as a condition of the surface slope.

Conversely to most of the investigations on storm surges, which are based on linearized equations, Hansen (1956, 1962) also took into consideration the non-linear effects. There is little doubt that non-linear forms of the two-dimensional, vertically intergrated hydrodynamic equations are in many respects preferable to the linearized forms. Miyazaki (1965) in his research work on a hurricane surge in the Gulf of Mexico applied the linearized equation, but introduced, in addition, a term of bottom friction, which was a combination of the square of the current velocity in the middle depth and the wind stress. However, as soon as non-linear terms are included into the equations, there arises the interaction between storm surge and tide. This implies that in sea regions with pronounced tides, the computations of storm surges without taking into account the latter phenomenon are not fully correct on the basis of these equations.

The problem of the interaction between storm surge and tide as based on the solution of hydrodynamic equations is non-linear. In shallow waters and estuaries effects representing the formation of shallow-water tides and the influence of the storm surge upon the tide must therefore be taken into account. Investigations of this type have, for instance, been performed by Proudman (1955a,b, 1957, 1958), Doodson (1956) and Rossiter (1961). These authors have examined the propagation of tide and storm surge in a narrow channel on the basis of one-dimensional equations of continuity and motion. These equations are:

$$\frac{\partial}{\partial x}\left[(h+\xi)\,Bu\right] + B\frac{\partial \xi}{\partial t} = 0 , \qquad \frac{\partial u}{\partial t} + u\frac{\partial u}{\partial x} = -g\frac{\partial \xi}{\partial x} - \frac{F_b}{\rho_w(h+\xi)}$$

where bottom fricton:

$$F_b = k\,\rho_w\,|u|u$$

In these equations x is the distance along the channel, h, as above, the average depth of the water, B the breadth of the channel for the average sea level, u the average value over a cross-section of the current in the direction of increasing x and k a coefficient representing friction. t, ξ, g and ρ_w are symbols for the same elements as concerned in the eq. 3–5. In the case of an estuary with varying dimensions, B and h are functions of x, while ξ and u are functions not only of x, but also of t.

Proudman (1955a) computed, with the help of the equations given above, the result of the propogation of tide and storm surge in an infinitely long estuary. Assuming that the disturbance was generated in the open sea and subsequently penetrated into the estuary, the procedure used by Proudman initially ignored the non-linear terms resulting in a solution which could be regarded as a first approximation to the final solution. Thereafter the non-linearity was taken into consideration as a second approximation and the final solution reached in this way. The results showed that, for simular meteorological conditions prevailing over the sea, the height of the storm surge in the estuary with a maximum occurring close to the time of the tidal high water is less pronounced than the height of the storm surge whose maximum occurs fairly close to the time of tidal low water. In addition, the height of the storm surge decreases with increasing tidal height. The results achieved by Proudman also indicate that, if the storm surge reaches its maximum at the approaches to the estuary only shortly before the arrival of tidal high water, and remains more or less unchanged until after tidal high water, the storm surge in the estuary may reach not less than two crests, the first before and the second after the time of tidal high water. According to Proudman this development is brought about by friction. In this connection we may refer to the numerical example illustrating the development of the storm surge of 1953 in the Thames estuary and reproduced in the following pages (pp. 77–80). This storm surge showed two distinct peaks.

In a later study, Proudman (1957) chose a uniform estuary of finite length with one end open to the sea and the other closed. The main purpose of the model was the investigation of the reflection at the closed end of the progressive wave moving inwards from the sea. The wave was suppposed to represent a combination of tide and storm surge. The method of successive approximations referred to above was used. If the estuary is short, it could be shown that the effect of the interaction between tide and storm surge results in a higher sea level, at the approaches to the estuary, if the maximum occurs at the time of tidal high water rather than at the time of tidal low water. This effect is due to the shallow-water terms in the equation rather than to friction.

Rossiter (1961) considered a channel of constant depth, but varying breadth, representing the Thames, and determined the elevation of the sea level along the channel for different combinations of storm surge and tide penetrating into the channel from the open sea. Rossiter's results showed that the effect of storm surge increases when it coincides with the rising tide, an interaction effect which could be confirmed for the Thames. According to Rossiter a negative storm surge has a retarding influence upon the tide, while a positive surge accelerates the progression. This is due to the fact that a negative storm surge reduces the depth causing a reduction in speed of the free wave. In addition, there is an increase in the effect of bottom friction, which also contributes to the retardation of the wave. A positive storm surge has the opposite effect. Assuming that the storm surge changes the phase of the tide, a simple explanation for the increase of the range of the storm surge height on rising tide follows.

Due to the practical significance of storm surges in the Thames estuary, this region has been investigated fairly extensively. The progress of the storm surges in other coastal waters in the North Sea, including the question of how a surge travelling along the east coast of Great Britain is modified by the influence of the tide in amplitude and time before reaching the head of the Thames, must be a task for further investigations. —

The North Sea is especially liable to develop storm surges in connection with exceptional weather conditions. These surges usually also affect the English Channel to some extent. Nevertheless, the consequences are generally more disastrous on the North Sea coasts, since the abnormally high sea levels flood the low-lying land in Lincolnshire and Essex in Great Britain and in The Netherlands.

Although the origins of storm surges and tides have nothing in common their behaviour is to some degree similar, since they progress in an anticlockwise direction around the North Sea. Fortunately, the maximum of the largest storm surges do not, as a rule, coincide with the time of the tidal high water, reducing in this way the otherwise still more destructive effects of the storm surge. It may be of interest to give in this connection a more comprehensive account of one of the most devastating storm surges which has occurred in the North Sea area: that of January 31 to February 1, 1953. The contributions of this storm surge to the sea level amounted to 270–330 cm in the southern parts of the North Sea. These disturbances were not the largest which have been recorded in this area, but at Southend at least the departure appears to be the highest which has occurred within an hour either side of the predicted high water.

The general meteorological situation is of the greatest significance in the generating of storm surges. In the North Sea area these surges usually occur when an exceptionally deep depression from the region north of Scotland progresses towards the east or southeast across the North Sea. When the centre of such a depression entered the North Sea during the morning of January 31, it was 968 mbar deep. Simultaneously a strong ridge with a high pressure of 1,033 mbar arose behind the depression. The consequence was gale-force northerly winds blowing over a considerable fetch. The winds themselves caused much damage in northern Scotland, but more fateful was the consequence that the tractive force of these winds began to drag a large quantity of water into the North Sea. By noon of January 31 the average sea level in the North Sea had started to rise considerably. It has been estimated that approximately $4.1 \cdot 10^{11}$ m^3 of water entered the North Sea between Scotland and Norway from 21h00 on January 31 to noon on February 1. These water masses raised the sea level of the whole North Sea by more than 60 cm above the level of high-water tide during most of this period. In Table XIII are given, according to Rossiter (1954b), but converted into cm, the contribution of the storm surge to the sea level for a great number of North Sea stations during January 31 and February 1, 1953. The values have been determined by deducting the predicted tide from the recorded sea-level data.

The effect of the Coriolis force upon the currents forced most of the water to follow the west coast of the North Sea. A slight part of the surplus water masses found an escape from the North Sea through the Strait of Dover. Rossiter estimated that roughly $0.48 \cdot 10^{10}$ m^3 of water per hour escaped southwards through the strait. Although this water quantity is only slightly more than one percent of the total water amount which had entered the North Sea from the north, it was sufficient to allow a decrease of the sea level in the southern part of the sea area. Rossiter was able to estimate that the escape of the water through the strait caused the sea-level heights in the region between the Thames estuary and Belgium to decrease at a rate of about 25 cm per hour. In spite of the approximate character of this determination, there is no doubt that the afflux of water through the Strait of Dover during a period characterized by a markedly raised sea level in the North Sea is a factor which must always be taken into account. In other words the strait acts in these cases as a minor safety valve.

In addition to the water masses which entered the North Sea, the sea level reacted to the effect of the low atmospheric pressure which contributed by further increasing the height. Finally, a marked oscillation was generated in the North Sea by resonance, moving along the coast of Great Britain and increasing continuously in amplitude as far as the Flemish Bight.

The data in Table XIII make it possible to reconstruct the progress of the storm surge. The crest of the surge reached Aberdeen between 15h00 and 18h00 on January 31, and by 18h00 the sea level was practically 100 cm higher than that predicted at Leith. By midnight the peak of the surge was rapidly approaching the coast of southern Holland.

TABLE XIII

THE CONTRIBUTION (CM) OF THE STORM SURGE TO THE SEA LEVEL IN THE NORTH SEA BETWEEN JANUARY 31 AND FEBRUARY 1, 1953 (ROSSITER, 1954b)

Hours (G.M.T.)	January 31								February 1							
	0	3	6	9	12	15	18	21	0	3	6	9	12	15	18	21
Aberdeen	-6	-6	-3	3	18	43	49	34	24	21	24	21	18	9	1	-6
Leith	9	6	6	9	21	56	98	95	73	58	61	61	55	46	37	27
R. Tyne entrance	-9	-12	-9	0	18	79	137	137	79	70	70	67	58	46	34	21
Hartlepool	-5	-3	-3	6	21	79	143	152	107	86	76	76	73	52	37	34
R. Tees entrance	6	9	9	21	49	110	174	220	168	125	128	143	125	101	85	73
Hull	-9	-21	-27	-31	0	58	122	180	183	137	101	82	85	115	113	52
Immingham	-27	-40	-37	-21	13	55	113	186	204	134	98	95	98	98	73	55
Grimsby	-18	-37	-31	-12	12	58	116	195	207	143	95	92	110	110	67	37
King's Lynn	-27	-15	3	21	31	81	183	271	262	198	149	149	156	143	128	116
Southend	-18	-34	-43	-34	-12	6	12	141	220	183	223	210	159	131	119	131
London Bridge	-24	-34	-37	-27	-27	-24	-21	0	207	140	143	137	183	140	98	95
Dover	-9	-6	3	9	24	37	61	116	186	189	183	152	137	122	107	85
Ostend	-12	-9	3	27	61	82	116	198	238	220	210	198	159	131	116	104
Brouwershavn	-24	-24	-9	27	79	110	143	262	311	284	230	259	223	156	131	131
IJmuiden	-12	-15	-12	18	67	104	140	229	268	281	262	192	177	143	104	73
Harlingen	-15	-31	-24	-15	15	73	186	238	180	311	339	204	116	122	128	101
Borkum	0	-9	-9	0	24	70	125	149	134	213	238	180	131	125	119	88
Norderney	0	-3	-6	-9	6	70	125	143	128	183	232	198	137	122	122	91
Cuxhaven	-3	0	-6	-15	-15	27	73	116	107	128	189	201	137	82	95	85
Husum	-6	-9	-9	-12	-12	31	91	110	91	101	156	152	110	85	85	58
List	9	21	27	21	3	-3	61	104	104	107	143	180	159	107	76	61
Esbjerg	-6	15	40	24	-15	-15	31	79	88	82	82	140	128	91	58	34
Hanstholm	6	9	9	3	-6	-12	-15	-6	3	6	-6	0	61	76	70	52
Hirtshals	8	6	6	-3	-15	-21	-24	-21	-21	-24	-24	-12	21	46	49	46
Nevlunghavn	0	0	-3	-9	-21	-27	-37	-40	-40	-40	-43	-37	-9	12	15	18
Tredge	0	0	-6	-9	-18	-24	-31	-37	-37	-37	-37	-24	-6	12	21	24
Stavanger	-15	-15	-18	-21	-24	-34	-43	-49	-45	-43	-37	-18	10	18	21	21
Bergen	-15	-18	-18	-12	-9	-21	-40	-52	-49	-40	-31	-18	-3	12	21	21
Maloy	-3	-3	0	-3	-12	-34	-67	-91	-58	-15	-21	-27	-21	-6	-3	-15

The sea-level heights, when most pronounced, were 200–250 cm above those predicted for southern Lincolnshire and the corresponding elevation was 300 cm and more for the Dutch ports. Progressing to the north and east along the coasts of The Netherlands, Germany and Denmark, the crest of the storm surge reached IJmuiden at about 03h00, Harlingen some two or three hours later, Esbjerg in Denmark at 10h00 and, finally, the southern parts of Norway at 22h00 on February 1. By this time the storm-surge character of the progressing water masses was practically extinguished. The storm surge and the tidal wave did not proceed at the same speed along the British coast, but at numerous localities the time of the occurrence of the maximum sea level was within one hour of the time of the predicted tidal high water. Rossiter, for instance, has given the following data characteristic of the sea-level heights between the Humber and Thames on January 31. These sea-level data are referred to Ordnance Datum (Newlyn):

	Imming-ham	Great Yarmouth	Lowe-stoft	South-end
Observed time of high water	19h00	22h04	23h10	24h45
Predicted time of high water	19h26	22h53	23h10	25h42
Observed height of high water (cm)	467	329	345	461
Mean high water springs (cm)	326	98	98	284

The storm surge occurred at the time of spring tides. However, if the surge had occurred a fortnight later, when one of the highest tides of the year was predicted, the damage would have been still more disastrous. Many areas were even then severely affected. In the Thames estuary two peaks were observed. It took approximately one hour for the first of these peaks to move from Southend to London Bridge. The reduction in height for this peak was only about 5 cm. The second peak took 5 hours to travel the same distance during which time it was reduced in height by 45 cm.

In addition to the comprehensive investigations of the storm surge of January–February, 1953, performed by Rossiter as quoted above, a considerable number of other studies have appeared on this subject. It may suffice to mention in this connection the following researches, all of which were published within a few years after the occurrence of the storm surge: Angeby (1953), Groen (1953, 1954), Guilcher (1953), Heyer and Grünewald (1953), Keuning (1953), Robinson et al. (1953), Steers (1953), Wemelsfelder (1953), Carruthers and Lawford (1954) and Lundbak (1955).

In all the cases discussed above the sea was ice-free. Let us now assume the special case where the sea is covered with fast ice. The wind stress at the sea surface is comparable with the wind stress at the surface of the open sea. The drag coefficient might, however, differ, since there is a difference in the roughness of the ice surface compared with that of the water surface. In addition, when the sea is completely covered by fast ice, the wind stress will be transferred to the coasts through the pressure exerted by the ice cover and no effects at all will be transferred to the water below the ice cover. The consequence will then be that the water will not be piled-up by the effect of the wind stress and only sea-level differences, due to the hydrostatic effect caused by the gradient of the atmospheric pressure corresponding to the wind, will be possible.

However, it is not very frequent that a sea such as the Baltic is completely covered by fast ice, for as a rule a very strong wind breaks the ice cover, producing drift ice and open water areas. In situations where some parts of the sea area of the Baltic are covered by fast ice and other parts are open or filled by drift ice, the slope of the water surface caused by wind stress must be expected to be in some degree reduced, depending on the extent of the area of fast ice. Since the shallow coastal regions are the first to be covered with fast ice, the deeper regions remaining open, it is very difficult to estimate the overall effect of the ice. It must always be kept in mind that the wind-stress-produced slope of the water surface is dependent on the sea depth.

In order to obtain data which are affected as markedly as possible by the occurrence of the fast ice cover, the northern part of the Gulf of Bothnia, the Bothnian Bay, was selected as the research area (Lisitzin, 1957a). Although the ice cover on the sea is generally not strictly continuous in this region, the amount of ice is as a rule considerable in cold winters, large ice floes more or less filling the entire basin. The main axis of the Bothnian Bay has a SSW–NNE direction. The length of the basin is approximately 300 km, the maximum breadth about 150 km and the average depth 40 m. The rather restricted extent of the Bothnian Bay makes probable the assumption, for the cases considered below, that the velocity of the wind does not deviate very markedly in different parts of the area. In the northernmost part of the Bothnian Bay is situated the sea-level station Kemi, while the corresponding station at Vaasa represents the southernmost part of the sea basin. Throughout the research period on which the results are based, four light vessels were in operation during the ice-free season. In addition, wind observations made on shore and on the islands were available from four meteorological stations.

From the wind observations made on board the light vessels during the 15-year period 1928–1942, a number of cases were selected for which the average wind force was at least 7 Beaufort (about 14 m/sec) and the wind direction more or less between S and SW. This procedure ensured that the piling-up of the water masses was accentuated, and hence the relative effects of other factors influencing the sea level and its variations considerably reduced. Moreover, the wind direction did not deviate much from the main axis of the sea basin. In order to ensure a more-or-less stationary slope of the water surface, the selection of the cases was restricted to those for which the velocity of the wind was relatively high and the wind directions deviated only slightly from those indicated above – at least for one observation prior to the one on which the computations were based. Finally, special attention was paid to the stipulation that the wind data from the various meteorological stations did not deviate very markedly from each other.

With the help of wind observations made on board light vessels, 40 cases could be selected which all corresponded to the above stipulations. For the three calm summer months, June to August, only one case a month could be utilized; in September, there were 6 cases; in October, 10 cases; and in November, not less than 14 cases. Since the light vessels, as a consequence of the increasing fast ice cover, are generally no longer in

TABLE XIV

THE RELATIONSHIP BETWEEN WIND VELOCITY AND SEA LEVEL

Date		Average wind (m/sec)	Sea-level difference, Kemi−Vaasa (cm)	α	Remarks on ice
	January				
1930	8/9	SW 16	50	2.6	weak ice in the skerries
1930	10/11	SW 16	50	2.6	weak ice in the skerries
1930	19/20	SW 16	50	2.6	weak ice in the skerries
1932	18/19	SSW 17	62	2.8	fast ice in the skerries, drift ice on the sea
1933	3/4	S 17	59	2.7	practically ice-free
1933	9	S 15	44	2.6	practically ice-free
1935	3/4	S 20	62	2.1	fast ice in the skerries
1935	10	SW 17	58	2.7	fast ice in the skerries
1937	5	SW 20	76	2.5	fast ice in the skerries
1938	21/22	S 17	52	2.4	practically ice-free
	February				
1928	4	SSW 19	26	1.0	the Bothnian Bay, except for a lane, covered by ice
1929	1	SSW 18	29	1.2	fast ice in the skerries
1930	13/14	SW 16	35	1.8	fast ice in the skerries
1930	28	SW 16	26	1.3	fast ice in the skerries
1931	10/11	S 16	21	1.1	probably the whole Bothnian Bay covered by ice
1931	21	S 16	33	1.7	continuous drift ice in the north part of the sea area
1940	26/27	SW 14	8	0.5	7/8 of the sea area fast ice, 1/8 drift ice
	March				
1928	17/18	SW 19	28	1.0	probably the whole area covered by thin ice
1938	6	SW 14	16	1.1	fast ice at least in the skerries
1940	19	SW 15	3	0.2	fast ice in practically the entire sea area
1942	29	SW 20	7	0.2	fast ice in practically the entire sea area
	April				
1938	21	SW 14	31	2.1	fast ice at least in the skerries
1939	19/20	SW 14	20	1.4	fast ice at least in the skerries
1939	29	SW 15	30	1.8	fast ice at least in the skerries
	May				
1934	18/19	SW 14	43	2.9	some drift ice on the sea
	June				
1935	17	S 14	48	3.2	no ice

TABLE XIV (continued)

Date		Average wind (m/sec)	Sea-level difference, Kemi-Vaasa (cm)	α	Remarks on ice
	July				
1942	26	SW 14	48	3.2	no ice
	August				
1938	20	SW 17	65	3.0	no ice
	September				
1930	24	SSW 18	52	2.2	no ice
1932	15	SW 19	57	2.1	no ice
1937	28	S 14	56	3.8	no ice
1940	11	SSW 14	52	3.6	no ice
1942	1	S 15	54	3.2	no ice
1942	24	SW 17	61	2.8	no ice
	October				
1930	8/9	S 14	41	2.8	no ice
1931	2/3	S 20	97	3.2	no ice
1931	28	SSW 21	85	2.6	no ice
1933	10/11	S 17	44	2.1	no ice
1934	23	SSW 17	62	2.9	no ice
1935	31	S 16	59	3.1	no ice
1937	15/16	SSW 18	75	3.1	no ice
1938	22	SSW 14	45	3.1	no ice
1938	24	SSW 14	53	3.6	no ice
1940	11	SSW 14	45	3.1	no ice
	November				
1929	12/13	S 21	96	2.9	no ice
1930	8	SSW 15	49	2.9	no ice
1930	22	S 18	59	2.5	no ice
1931	21	SSW 14	56	3.8	no ice
1932	23	SSW 14	41	2.8	no ice
1932	29	SW 18	65	2.7	no ice
1934	1/2	SSW 21	106	3.2	no ice
1935	26/27	S 17	71	3.3	no ice
1937	25	SW 20	86	2.9	no ice
1938	27	SSW 16	54	2.8	no ice
1939	24	S 18	75	3.1	no ice
1940	11	S 14	56	3.8	no ice
1941	14	SW 15	52	3.1	no ice
1941	25/26	SSW 14	52	3.5	no ice
	December				
1928	24	SW 17	41	1.9	fast ice in the skerries
1929	8	S 14	49	3.3	practically ice-free
1929	26/27	S 14	37	2.5	weak ice in the skerries
1931	4	S 20	81	2.7	practically ice-free

TABLE XIV (continued)

Date		Average wind (m/sec)	Sea-level difference, Kemi-Vaasa (cm)	α	Remarks on ice
	December				
1932	18	SW 17	54	2.5	practically ice-free
1936	4/5	S 16	51	2.6	practically ice-free
1936	15	S 19	74	2.7	practically ice-free
1936	18	SSW 16	54	2.8	practically ice-free
1938	9/10	S 17	64	3.0	practically ice-free
1940	16/17	SSW 17	62	2.9	fast ice in the skerries

operation by December, the cases illustrating the winter proper had to be completed with wind observations made at the coastal and island stations. These observations, although less appropriate for this study than those made on board light vessels, are doubtless adequate enough, especially when the final results are based on the average of several cases. For the 15-year period the results are based on 10 cases for December and January, 7 cases for February, 4 cases for March, 3 cases for April and, finally, only one case for May. The number of cases reflects very distinctly the seasonal cycle of wind velocity in the research area.

The relevant results are set out in Table XIV. The first column in this table gives the dates of the cases which have been found suitable for the study. The second column gives the estimated average direction and velocity of the wind (in m/sec) over the whole area of the Bothnian Sea. In the third column are reproduced the differences in sea level between Kemi and Vaasa. The maximum differences generally occur a few hours later than the highest observed wind velocity and coincide, as a rule, with the maximum sea level at Kemi. The fourth column gives the values of the coefficient α in formula (2). Finally, in the last column are given brief remarks characterizing the corresponding ice situation. These remarks give a conception of the relationship between the value of the coefficient α and the extension of the ice cover.

A closer study of the values of the coefficient α for the different months reveals a considerable number of fairly pronounced differences. The principal feature, however, is quite distinct. The coefficient has a higher value during the ice-free period than in winter time and decreases with increasing extension of the ice. For the extreme cases, in March 1940 and 1941 when the fast ice cover extended practically over the whole area of the Bothnian Bay, the coefficient was 0.2 and in February 1940, with 7/8 of the basin covered with fast ice, it was 0.5. It may be pointed out that the hydrostatic effect of the atmospheric pressure is not taken into account in these cases.

In Table XV are given the average values for α for the different months and the corresponding mean deviations. In February and March, when the continuous ice cover is generally at its maximum and the ice, owing to its marked thickness, is strong enough to withstand the effect of the wind, the slope of the water surface is on average approxi-

TABLE XV

THE AVERAGE VALUES OF THE COEFFICIENT α AND ITS MEAN DEVIATIONS FOR DIFFERENT MONTHS

	J	F	M	A	M	J	J	A	S	O	N	D
Coefficient α	2.6	1.2	0.6	1.8	2.9	3.2	3.2	3.0	3.0	3.0	3.1	2.7
Mean deviation	0.2	0.4	0.5	–	–	–	–	–	0.7	0.4	0.4	0.4

mately three times weaker than during the ice-free period. April, December and January probably represent transitory stages.

The average value of the coefficient for the ice-free period is 3.0, i.e., somewhat less than the value given in formula (2). The deviation is not pronounced and may easily be accounted for by the over-estimation of wind velocity. Since the main purpose of the investigations was the relative effect of wind stress during the ice-free period on the one hand, and during winter months on the other hand, no effort has been made to correct the wind observations.

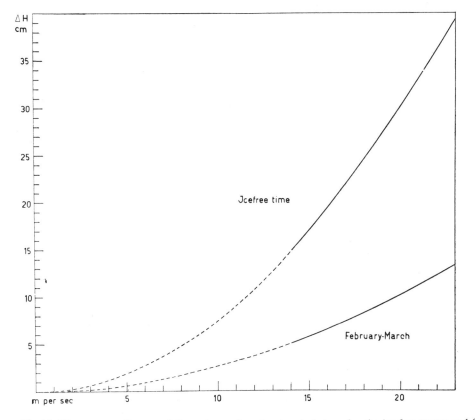

Fig. 17. The average piling-up of the water surface due to wind stress for the ice-free season and for the months February and March (Lisitzin, 1957a).

Fig. 17 represents the average piling-up of the water surface caused by wind stress for a distance of 100 km in the Bothnian Bay between Kemi and Vaasa both for the ice-free season and for the months February to March. The curves have been drawn for wind velocities between 0 and 23 m/sec, although the lower velocities (less than 14 m/sec) have not been taken into consideration in the study, since their influence on sea level may frequently be negligible in comparison with other factors which affect the sea surface. The deviation between the two curves is highly pronounced.

THE CONTRIBUTION OF WATER DENSITY

Density is one of the fundamental physical properties of sea water. It is a function of temperature, T, salinity, s, and pressure, p. The symbol for water density is $\rho_{s,T,p}$. The density of pure water is equal to unity and is at its highest at $4°C$, but in sea water the density differs from unity, not only because of the temperature effect but also as the consequence of the presence of salt which causes the density to exeed that of pure water by a small amount. The density of sea water thus decreases with changing temperature and by the dilution of the water due to the discharge of rivers, precipitation and melting of ice. On the contrary, the density increases as the result of increasing salinity, cooling down and warming up to $4°C$, evaporation and ice formation. Most of these processes occur at the sea surface. As soon as the density at the water surface becomes higher than in the deeper layers the vertical convection, the so-called thermo-haline convection, sets in.

The static pressure p at a depth h of the water column is determined by the force exerted by the water masses above acting on an area of 1 cm^2. If the atmospheric pressure which acts at the water surface is left out of consideration, the static pressure is determined by the formula:

$$p = \rho_{s,T,p} \cdot g \cdot h \text{ dyne/cm}^2$$

In this formula $\rho_{s,T,p}$ is the average water density in situ at the depth h and g is the acceleration of gravity. V. Bjerknes introduced the term decibar. 1 decibar (dbar) corresponds to a pressure of 10^5 dyne/cm^2. The pressure in decibars and the water depth in metres coincide with each other within the limits of one percent.

Let us assume that the density of sea water at a given locality and for a given time is 1,02575. This figure is rather cumbersome to handle and an abbreviation symbol has therefore been introduced:

$$\sigma_{s,T,p} = (\rho_{s,T,p} - 1) \cdot 1,000$$

The desity $\rho_{s,T,p} = 1,02575$ may thus be written $\sigma_{s,T,p} = 25.75$

The density of sea water in situ is an important parameter not only for the dynamic processes in the oceans, but also as one of considerable significance with regard to the study of variations in sea level. For this purpose it is, however, generally more convenient to use the reciprocal value of the density, i.e., the specific volume of the water in situ, $\alpha_{s,T,p}$. Different tables have been compiled and are available to facilitate the computations of the density and the specific volume of the water in situ.

The effect of water density upon the sea level was recognized long ago and the number of studies on this subject have been extremely prolific. Especially rewarding in this connection was the research work done by Pattullo et al. (1955). The method used by these authors and the principal results obtained by them will therefore be recapitulated briefly below.

From the available oceanographic data, information concerning variations in the temperature and salinity in the oceans and seas may be obtained for the particular months at various depths (or pressures). For any depth ΔT and Δs designate the monthly departures in temperature and salinity from their annual averages $\bar{\bar{T}}$ and $\bar{s}$. For small values of ΔT and Δs the corresponding departures in specific volume are given by:

$$\Delta\alpha = \alpha(T, s, p) - \alpha(\bar{T}, \bar{s}, p) = (\partial\alpha/\partial T)\,\Delta T + (\partial\alpha/\partial s)\,\Delta s + \dots$$

where $\partial\alpha/\partial T$ and $\partial\alpha/\partial s$ are to be evaluated at $\bar{T}$, $\bar{s}$, p. However, the explicit dependence of $\partial\alpha/\partial T$ and $\partial\alpha/\partial s$ on pressure has been neglected and accordingly in the numerical computations $\partial\alpha/\partial T$ and $\partial\alpha/\partial s$ have been read as functions of $\bar{T}(p), \bar{s}(p), 0$.

The 'thermal' and 'haline' departures are thus determined by:

$$z_T = g^{-1} \int_{p_a}^{p_0} (\partial\alpha/\partial T)\,\Delta T\mathrm{d}p, \qquad z_s = g^{-1} \int_{p_a}^{p_0} (\partial\alpha/\partial s)\,\Delta s\mathrm{d}p$$

and the added effect of the two terms, denominated by Pattullo et al. as steric departure, is:

$$z_a = g^{-1} \int_{p_a}^{p_0} \Delta\alpha\mathrm{d}p$$

Here p_a is the atmospheric pressure and p_0 the pressure to which the integration has to be carried out, presumably the pressure where all seasonal effects vanish.

The results achieved by Pattullo et al. are fairly interesting. The steric sea level is practically equal to the sum of the thermal and haline departures. These two types of departure have, however, quite different characteristics.

The average range of thermal departures was, according to the extensive data on which the authors have based their computations, 11 cm. Small thermal departures were observed in some equatorial and in most polar regions. The maximum ranges, reaching approximately 25 cm, have been observed in the Sea of Japan and in the regions situated to the north of Bermuda. All these results are based on monthly means, not on particular observations.

The main characteristic of haline departures was that they did not show any more pronounced tendency to cluster around the average value. Two-thirds of the z_s computations resulted in negligible ranges of 5 cm or less. The haline departures were especially weak around Bermuda and in the extensive oceanic areas situated to the east and southeast of Japan. Pronounced values of haline departures have been observed for one location in the Bay of Bengal, reaching a range of 41 cm and along the continental slope off eastern Asia in the region extending from Formosa to Hokkaido.

On the whole, the steric departures are more or less in phase with the recorded departures. The considerable irregularities occurring from one place to another, do not normally allow any easy generalization. Some characteristic features may, however, be mentioned.

In the equatorial regions the steric departures are mainly thermal and relatively slight, while in the subtropical latitudes the steric departures, being mostly thermal, are fairly large. On average they seem to agree rather well with the recorded sea-level heights.

In the subpolar latitudes the thermal departures are small, while the sea-level variations, as a rule, reach marked values. The contribution of salinity is difficult to determine in these regions. Where observations on salinity are available, the depth of the water is generally slight. Deep-water salinity data are available to a greater extent only from the Oyashio area. In this region the haline departures are equivalent to the thermal departures, and together they result in a fairly good agreement between the steric departures and the variations in the recorded sea level. The data used by Pattullo et al. in their investigations are assuredly too restricted in their quantity to allow more general conclusions. Nevertheless, they show that water density is only one of the factors affecting the fluctuations in sea level. A more detailed study, based on still more extensive data, was therefore necessary.

Such studies have been performed, for instance, for the Mediterranean and the Baltic (Lisitzin, 1959a,c). In the Mediterranean a correlation between the seasonal cycle in sea-level variation and that of the changes in water density could be expected according to the results of Pattullo et al. Reproducing the density data given by these authors as the average of the results covering an area situated to the southeast of southern Italy and Sicily, and completing them with the values for two oceanographic stations, St.I and St.II, located at 2.5 km and 6.3 km respectively outside Monaco, we obtain the following series:

J	F	M	A	M	J		J	A	S	O	N	D

Data given by Pattullo et al.:

−3.0	−4.4	−5.3	−5.4	−2.0	−0.2		3.2	3.4	2.3	5.2	5.8	0.4

Station St. I

−2.7	−5.0	−5.4	−2.9	−0.7	1.2		1.4	3.7	4.0	4.4	2.1	0.1

Station St. II

−2.3	−4.9	−5.3	−3.7	−1.3	0.3		1.0	4.9	4.7	4.4	2.3	0.2

The three series show a considerable correspondence, although the maximum occurs for the data given by Pattullo et al. one month later than for St.I and three months later than for St.II. However, if the harmonic constants are computed for the three series of data, the analogy between them becomes still more accentuated. The results are in this case:

	Pattullo et al.	Station St. I	Station St.II
Amplitude	5.2 cm	4.5 cm	4.5 cm
Phase	181°	169°	173°
Date of maximum	September 22	September 10	September 14

All the data indicate that the maximum sea level due to changes in water density occurs in the middle of September.

Turning to sea-level records the following harmonic constants may be given (Lisitzin, 1954b):

	Amplitude (cm)	Phase	Date of maximum
Marseilles	3.9	235°	November 16
Monaco	4.0	263°	December 14
Porto Maurizio	3.6	202°	October 13
Civitavecchia	3.6	238°	November 19
Cagliari	6.0	180°	September 21
Palermo	4.1	227°	November 8
Average	3.7	221°	November 2

Since the recorded sea level reaches, on average, a maximum at the beginning of November, there is — with the exception of Cagliari — a distinct retardation of the seasonal cycle in sea level compared to that of the density of the water. This retardation is too marked to be attributed to occasional disturbances. The results show that in addition to the contribution of density other terms must be taken into account. Since the deviations between the results based on water density and those referring to the sea-level records are most pronounced for the stations situated along the continental coast, the wind effect is probably the main factor contributing to the deviations.

In the Baltic Sea the influence of the variations in water density upon the total fluctuations of sea level is fairly slight. Water density contributes, in fact, only some 10% of the range of the monthly mean values, which, for instance, varies along the Finnish coast between the limits of 24 and 33 cm. The data given in Table XVI refer to the first day of each month for the hydrographic station Märket which is the deepest of the Finnish stations at which regular observations are made throughout the whole year. In order to obtain a concept of the effect of the depth upon which the sea-level heights computed from water density are based, the results are given for every ten metres from 10 to 100 m depth (Lisitzin, 1959c).

Table XVI shows that the range of variations reaches a maximum for the depths of 30 and 40 m, being 2.6 cm at these depths. It has already been mentioned above that the variations in sea level are only in slight measure dependent on the fluctuations in water density. It may nevertheless be of interest to examine the way in which elimination of the effect of density variations influences the residue of sea-level data. For this purpose the

TABLE XVI

VARIATIONS IN SEA LEVEL (CM) AT MÄRKET, DUE TO THE FLUCTUATIONS IN WATER DENSITY, FOR THE FIRST DAY OF EVERY MONTH AND FOR DIFFERENT DEPTHS

Depth (m)	J	F	M	A	M	J	J	A	S	O	N	D
10	−0.3	−0.4	−0.4	−0.3	−0.2	−0.2	0.3	1.0	0.9	0.2	−0.2	−0.3
20	−0.6	−0.7	−0.6	−0.4	−0.4	−0.4	0.4	1.6	1.5	0.4	−0.4	−0.5
30	−0.7	−0.8	−0.7	−0.5	−0.5	−0.4	0.4	1.8	1.8	0.5	−0.3	−0.6
40	−0.7	−0.8	−0.7	−0.5	−0.5	−0.5	0.3	1.7	1.8	0.6	−0.2	−0.5
50	−0.5	−0.7	−0.6	−0.4	−0.5	−0.5	0.5	1.6	1.6	0.7	0.0	−0.3
60	−0.5	−0.7	−0.6	−0.4	−0.6	−0.7	0.1	1.3	1.4	0.6	0.0	−0.2
70	−0.4	−0.6	−0.5	−0.4	−0.6	−0.7	0.0	1.2	1.3	0.5	0.2	0.0
80	−0.3	−0.5	−0.4	−0.4	−0.5	−0.7	−0.1	1.0	1.1	0.4	0.2	0.2
90	−0.2	−0.5	−0.4	−0.4	−0.5	−0.7	−0.1	1.0	0.9	0.3	0.2	0.2
100	−0.2	−0.4	−0.3	−0.2	−0.5	−0.7	−0.1	0.9	0.8	0.2	0.2	0.2

harmonic constants of the annual and semi-annual constituents for Hanko (Hangö) were determined on the basis of the monthly mean values for the period 1888–1957, using on the one hand the original data and on the other hand those corrected for the density effect: The results were, in the former case:

Sa: range 10.0 cm, phase 216°,
Ssa: range 4.3 cm, phase 249°,

and in the latter cese:

Sa: range 9.7 cm, phase 211°,
Ssa: range 4.1 cm, phase 248°.

In the latter case the sum of the ranges has decreased by 0.5 cm and the maximum of the annual constituent occurs 5 days earlier. The elimination of the density effect hence does not appreciably influence the harmonic constants. This fact shows once more that the contribution of the density term is practically negligible for the variation in sea level in the Baltic.

THE EFFECT OF CURRENTS

As a consequence of the uneven distribution of heat over the Earth's surface, differences in the atmospheric pressure and water density arise between the equator and the poles. The gravitational force tends to level out these differences, which results in the circulation of the air and water masses. The gravitational force is the added effect of the attractive force of the Earth and the centrifugal force of the Earth's rotation. The tide-generating force, being much weaker than the two forces mentioned above, may be left out of consideration in this connection. The acceleration of the gravitational force, or gravity, may be computed on the basis of the 'international gravity formula' accepted in 1930. For the sea surface this formula reads:

$$g_0 = 9.78049\,(1 + 0.005288\,\sin^2\varphi - 0.000006\,\sin^2 2\varphi)\ \text{m/sec}^2$$

Gravity is thus a function of the latitude φ reaching the highest values at the poles and the lowest at the equator. Moreover, gravity is dependent upon the distance from the sea surface, increasing with increasing depth. This incresse is, however, extremely weak, for instance at a depth of 1,000 m amounting to only 0.2‰ of the value at the sea surface.

Since gravity is the most significant force it is necessary to build up a coordinate system where this fact is taken into consideration, basing the system not on surfaces of equal water depth but on surfaces of equal gravity, called potential surfaces. These surfaces are everywhere perpendicular to the direction of the gravitational force. No work is required for the displacement of mass along the potential surface if no force besides gravity acts on this surface. The work which must be done to move a mass unit the distance h (in metres) in the vertical direction is determined by the potential value gh. This potential may also be used to characterize the position of any potential surface. V. Bjerknes introduced the term dynamic decimetre as the unit for potential, corresponding to the work required to move a mass unit approximately one decimetre. Starting from the ideal sea-surface level corresponding to the potential value zero as a reference surface, the potential surfaces may thus be defined as surfaces of equal dynamic depth. Choosing the dynamic metre as a unit for the dynamic depth we obtain $D = gh/10$. The unit of gravity potential of 1 dynamic metre corresponds roughly to 1.02 geometrical metres. In spite of the close numerical agreement between the two units, it must always be remembered that the dynamic metre represents work and has the dimensions of work. A few numerical examples of the correspondence between the geometrical depths and the dynamic depths are given below:

Geometrical depth (m)	Dynamic depth (dyn. m)
100	98
500	490
1,000	980
5,000	4,903
10,000	9,811

The atmospheric circulation, due to the uneven distribution of air masses, is more pronounced than water circulation in the oceans and seas, which is a consequence of the fact that air density is considerably less than water density. Nevertheless, as soon as the air currents moving over the water surface reach a higher velocity, there arise at the boundary between the two elements, as the result of the frictional effect, so-called wind or drift currents. These currents are of decisive significance for the water circulation in the oceans and seas. They exert an additional effect on the distribution of the water masses.

For an understanding of the character of the ocean currents, it must always be taken into account that every motion occurring on the rotating Earth's globe is influenced by the deflecting force of the Earth's rotation. The equation representing this force, generally called the Coriolis parameter, reads:

$$f = 2 \omega v \sin \varphi$$

where v is the velocity of the current, ω the angular speed of the Earth's rotation – amounting to $2\pi/86,164$ sec $= 7.29 \cdot 10^{-5}$ sec^{-1} – and φ the geographical latitude.

The Coriolis force is only secondary in character. In analogy with the centrifugal force it is not able to bring about a motion, but arises exclusively as the consequence of an already existing movement. The effect of the Coriolis parameter is noted by the deflection of the mass unit to the right in the northern hemisphere and to the left in the southern hemisphere, when looking in the direction of the current, in this way causing differences in sea level. Since the expressions right and left may in some cases cause confusion, it may be more appropriate to speak of a motion 'cum sole' and 'contra solem', i.e., with or against the apparent azimuth motion of the Sun in the equatorial plane.

In 'order to simplify the problem connected with the currents and the heights of the sea level, the average circulation in the oceans is considered to involve only two forces, the gravitational and the Coriolis forces. In this case the surface of the world oceans represents a level which is perpendicular to the resultant of the two forces. Since the Coriolis force is only a small fraction of the gravitational force, the surface of the oceans, at least as long as the velocity of the current is not marked, deviates only slightly from the surface of a geoid.

If the heights of sea level at two places situated at a distance L from each other are h_1 and h_2, the tangent of the angle β which characterizes the slope of the water surface arising from the current motion may be expressed in the following way:

$$\tan \beta = \frac{h_1 - h_2}{L} = \frac{f}{g} = \frac{2 \omega v \sin \varphi}{g}$$

or:

$$v = \frac{(h_1 - h_2) g}{2 \omega \sin \varphi \cdot L} \tag{1}$$

which indicates that the current velocity is proportional to the difference in the height of the sea level on two sides of the current at constant latitude.

The next step is to connect the velocity of the current with the density or the specific volume of the water. The hydrostatic pressure p at a given depth h is determined by the force exerted by the weight of the water volume per unit area. We have thus:

$$p = \bar{\rho} \, gh$$

where $\bar{\rho}$ is the average density of the water volume in situ. The term gh may be expressed in dynamic metres and one thus obtains $p = \bar{\rho}D$. The pressure exerted by a water column of the height corresponding to 1 dyn. m is called 1 decibar (dbar, see p.86) and used as a pressure unit. 10 dbar = 1 bar and correspond approximately to 1 atmosphere, which is used as a unit in engineering.

Assuming that $\bar{\rho}_1$ and $\bar{\rho}_2$ are the average values of water density corresponding to the depth h_1 and h_2, and $\bar{\alpha}_1$ and $\bar{\alpha}_2$ are the corresponding values of the average specific volumes, the pressure p (in dbar) at the concerned points is:

$$p = \bar{\rho}_1 g h_1 = \bar{\rho}_2 g h_2$$

if the sea-level heights are determined starting from an isobaric surface, where, in addition, no noteworthy current occurs. Thus:

$$h_1 = \frac{p}{\bar{\rho}_1 g} = \frac{\bar{\alpha}_1 p}{g},$$

$$h_2 = \frac{p}{\bar{\rho}_2 g} = \frac{\bar{\alpha}_2 p}{g}$$

Finally we obtain:

$$v = \frac{p(\bar{\alpha}_1 - \bar{\alpha}_2)}{2\omega \sin \varphi \cdot L} = \frac{D_1 - D_2}{2\omega \sin \varphi \cdot L}$$

The relative internal pressure field in the oceans may be computed as soon as the vertical distribution of temperature and salinity has been determined for a sufficient number of points in the area under consideration. The relative dynamic topography is then represented by a number of lines of equal dynamic height, which are also sometimes called the dynamic isobars. Pattullo et al. (1955) introduced the term 'steric' sea level to denote the dynamic heights. This was done in a publication dealing with the seasonal oscillation of the mean sea level. The reason for the introduction of the new term was mainly the fact that the dynamic heights have the dimension of work and not of height, which should be the basis for sea-level changes.

A few additional words must be devoted here to the distribution in situ of water density and its reciprocal value, the specific volume. It has already been mentioned above (p. 86) that the two elements are dependent on temperature, salinity and pressure. Starting from a water column with the temperature $T = 0°C$ and the salinity $s = 35‰$, corresponding to the specific volume denoted by the symbol $\alpha_{35,0,p}$, it is possible, with the help of a number of necessary tables, to compute the anomaly δ of the concerned specific volume in situ $\alpha_{s,T,p}$, using the formula:

$$\alpha_{s,T,p} = \alpha_{35,0,p} + \delta$$

According to a simplified method developed by H.U. Sverdrup in 1933 the anomaly of the specific volume consists of three terms, the first depending only upon temperature and salinity, the second on salinity and pressure and the third on temperature and pressure. The necessary tables for the simple, if somewhat time-consuming, computations of these terms were given by Sverdrup et al. (1942) and Dietrich (1952). The use of computers facilitates the work considerably.

The above-outlined dynamic computations have found extensive application in connection with the determination of the mean sea level in the different oceans and their particular parts. In this respect it may be sufficient to refer to the chapter 'A World-wide Mean Sea Level and its Deviations'.

We will now proceed to consider more closely the effect of the deflecting force of the Earth's rotation in terms of sea-level differences in some special cases. Let us assume that over an elongated but comparatively narrow bay in the northern hemisphere, the wind is blowing from the sea in the direction of the axis of the basin. If the wind is strong and persistent, a drift current of considerable strength arises, transporting water towards the inner parts of the bay. The Coriolis force is continuously acting on the current, bringing about a difference in sea level between the right-hand and the left-hand coasts. As soon as the wind effect has ceased, the water masses piled up in the inner parts of the basin begin to stream backwards, or to express it in other terms, the current changes its direction and the consequence is that the difference in sea level at the opposite coasts gradually decreases and, finally, changes sign. This phenomenon has frequently been observed and examined in the extensive bays bordering Finland. Reference may, for instance, be made to the paper published by Palmén and Laurila (1938), which gives an insight into the mechanism of the processes and the stratification of the water which occur in this connection.

Next, the effect of atmospheric pressure on the sea level in the form of wind and its subsequently current-producing factor will be considered (Lisitzin, 1964b). Every atmospheric disturbance over the oceans causes, in addition to the hydrostatic effect, air motion which produces currents. These currents may reach the final phase of their development only if the duration of the original disturbance is long enough to assure this development. If the currents which are the consequence of persistent high- and low-pressure centres in the deep open parts of the oceans are considered, it seems highly probable that the yearly mean values, in spite of considerable seasonal deviations, may give a general picture of the phenomenon as a whole.

The theoretical computations are based on the following equations:

(1) The equation giving the velocity of the geostrophic wind W is a function of the difference in atmospheric pressure p, the air density ρ_a, the distance L and the geographical latitude φ:

$$W = \frac{\Delta p}{\rho_a \, 2\omega \sin \varphi \cdot L}$$

(2) The relationship between the velocity of the wind at the sea surface w and the velocity of the geostrophic wind W:

$$\frac{w}{W} = \frac{\sin \alpha}{1 + \dfrac{k \sin \gamma}{2\omega \sin \varphi}}$$

where α is the angle deviation, k the frictional coefficient and γ the friction angle over the sea.

(3) The empiral formula giving the ratio between the current velocity v in cm/sec and the velocity of the wind w in m/sec:

$$v = 1.2 \, w$$

(4) The equation representing the effect of the Coriolis force upon the sea level ΔH:

$$\Delta H = \frac{2\omega \, v \sin \varphi \cdot L}{g}$$

On the basis of these equations it is possible to determine the slope of the water surface in a region with high or low atmospheric pressure. Since the exact values of the elements involved in the computations are not known, the results are only approximate. They indicate, however, that the final interaction of the different terms in the system: atmospheric pressure—wind—current—sea level is considerably more pronounced and opposite in sign than the hydrostatic effect of the atmospheric pressure upon the sea level. In order to give a concept of the magnitude of the results achieved theoretically, it may be mentioned that an increase of 1 mbar per 100 km in the atmospheric pressure corresponds to an increase of sea level, due to the Coriolis force as given above, of the following magnitude:

Latitude	Increase in sea level, due to Coriolis force (cm)
20°	4.5
40°	6.2
60°	6.9
80°	7.1

According to these results the increase in sea level is on average 6 cm, while Zubov (1959), for instance, using a different approach to the problem, has estimated it at 10 cm.

Fortunately, there is a possibility of estimating the validity of the theoretical result by studying the distribution of the dynamic heights in the oceanic regions characterized by high or low atmospheric pressure. The results obtained in this way are also only approximate, since factors other than the local atmospheric disturbance influence the dynamic topography of the water surface. In addition, it must always be kept in mind that the data on which the computations of the dynamic heights are based are, as a rule, not synoptic.

A comparison of the values for atmospheric pressure with the results given by Defant (1941) for the dynamic topography in the latitudes of 45°N and 45°S in the Atlantic

Ocean leads to the following average relationship: the departures in sea level in cm correspond to 6.2 times the deviations in atmospheric pressure in mbar. For the Pacific Ocean the estimate of the relationship is still more approximate; however, in this case also the estimate is close to 6.2.

By a similar procedure the concerned relationship can for the latitudes of 30° be determined at 5.5 and for the latitudes of 15° at 3.9. All these results are compared with the corresponding theoretical values. The mean deviations for the observed relationship give an idea of the reliability of the results:

Latitude	Computed theoretically (cm)	Determined on the basis of observations (cm)
15°	3.6	3.9 ± 1.1
30°	5.5	5.5 ± 1.5
45°	6.4	6.2 ± 0.6

The effect of atmospheric disturbance upon the inner stratification of the different water layers is clearly illustrated by the schematic picture reproduced in Fig 18. The drawing to the left shows that the high-pressure atmospheric centre is associated in the oceans, as a consequence of the anticyclonic circulation of the current, with a thick layer of water characterized by low density. On the contrary, in the case of a low-pressure centre, as shown by the drawing on the right in the figure, the layer of less dense water is extremely thin. In this connection it must be emphasized once more that the above results are valid only in cases of stationary high- and low-pressure areas, while the hydrostatic effect of atmospheric pressure is probably the predominant factor in all remaining situations prevailing over the oceans.

As a third illustration of the effect of the Coriolis parameter upon the currents and consequently also on the distribution of sea-level heights, a narrow transition area between two sea basins has been chosen. In the Danish Straits, for instance, this effect

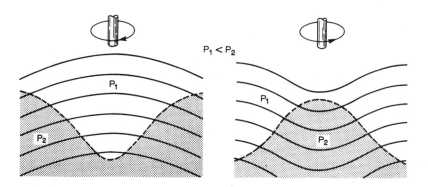

Fig. 18. The effect of the atmospheric disturbances upon the stratification of the water layers (Lisitzin, 1964b).

should be distinctly reflected in the differences in sea level on opposite coasts of the sounds. Sea-level records may therefore be used also, as an indicator of the in- and outflow of water to and from the Baltic Sea, and thus of the changes in mean sea level and the water balance in the basin as a whole. The knowledge of the water balance is without doubt a factor which is of greatest significance for the basic study of pollution in the Baltic, which is considered by some oceanographers as one of the most important problems of the area. It must, however, be pointed out that the positions of the recording gauges in the transition area around Denmark were originally selected on the basis of considerations which did not always correspond to those essential for the investigation of the effect of the Coriolis force upon the currents. Nevertheless, five pairs of sea-level stations have been continuously in operation for a long time on opposite coasts of various straits in the waters around the Danish coasts.

The question of the determination of the amount of the water transport among the particular sounds in the transition area as a consequence of the prevailing meteorological situation has hardly been studied on any larger scale. This problem is especially difficult in the cases where the research into the relationship between the slope of the water surface in the particular parts of the transition area and the water balance in the Baltic Sea are based on monthly mean data. On the other hand, it is not possible to examine this question for separate cases, principally due to the fact that sea-level data are, as a rule, not available for large coastal areas in the southern and eastern parts of the sea basin. The most reliable data so far for monthly sea-level variations in the entire Baltic Sea were computed by Hela (1944) for the 10-year period 1926—1935. These results were based on sea-level records made at 12 stations selected in such a manner as to be representative of all the separate parts of the Baltic basin. On the other hand, it may be pointed out that sea-level observations made at Utö, an island situated in the northern part of the Baltic proper, are generally sufficient to give a concept of the variations of the amount of water in the whole sea basin during a prolonged period, although they are not always character-istic of particular cases. In order to compare the data computed by Hela for the whole Baltic and the corresponding values for Utö, the monthly averages for the 10-year period in question were intercorrelated and resulted in a correlation factor of 0.991. The highly pronounced correlation makes reasonable the use of the monthly sea-level averages at Utö as representative of the whole sea basin. Since the observations have been taken at Utö almost continuously since 1865, there are at present data available for a time-span cover-ing more than 100 years. (These data are published in the series of the Institute of Marine Research in Finland, the *Merentutkimuslaitoksen Julkaisu/Havsforskningsinstitutets Skrift.*)

However, for a preliminary study of sea-level variations in the Baltic as a function of the differences of sea level in the transition area, the 10-year period 1926—1935 was selected (Lisitzin, 1972c). This study has shown clearly, that there exists in numerous cases a close connection between the two factors, but also indicated that the number of deviations is pronounced. The latter fact is due to the complexity of the relative distribu-tion of sea-level heights in the transition area, as a consequence of the complexity of the prevailing meteorological and hydrographic situation. Nevertheless, there appeared in the course of the work an interesting feature. During the months where the departures in sea level at Utö were considerable, there was also a very marked difference in sea-level heights between the two Danish stations Hornbaek and Gedser. It could therefore be concluded that this difference reflects better than the slope of the water surface in the particular straits the characteristics of the sea-level situation in the transition area. It seemed appropriate, therefore, to intercorrelate the differences in sea level at Hornbaek and Gedser with the sea-level changes at Utö for the years 1926—1935. This procedure resulted in the correlation factor $r = 0.849$ and the probable mean deviation $s = 0.017$. This high correlation allows the determination of the equation representing the relationship between the monthly sea-level differences at Hornbaek and Gedser (x) and the corresponding changes in sea level (y), observed at Utö. The result was:

$$y = 2.141 \, x$$

Other practical combinations of stations and station groups in the transition area provided less satisfactory results. For instance, the intercorrelation of sea-level differences at Warnemünde and Gedser with the sea-level variations at Utö resulted in the correlation factor $r = 0.538$.

The high correlation for Hornbaek and Gedser encouraged us to proceed with research in the same direction but extending it over a more prolonged period. More-or-less complete data being available for the two stations since the year 1892, the 75-year period extending from this year to 1966 was selected. Owing to the more prolonged period, all sea-level data were corrected for the effect of the secular variation in mean sea level. In this case the results were $r = 0.797$, and:

$$y = 1.945 \, x$$

Although the correlation factor shows a slightly lower value for the longer period than for the shorter one, it may be mentioned that the probable mean deviation was in the former case approximately half of the latter case. In addition, since the values of the ratio between y and x differ from each other by less than 10%, they produce a result which is acceptable for all practical purposes. This result may then be of use in connection with the study of the water balance in the Baltic Sea, although the contribution of other factors — river discharge, precipitation and evaporation — must always be kept in mind.

Current measurements are generally considered to be more time-wasting than sea-level observations and in all the cases where automatic current meters are employed, the continuous care and supervision require considerable effort. The possibility of building up a scheme of the average directions of the current system in a sea basin may therefore be of great interest, even if this scheme is only approximate. An example of such a scheme is given for the Ligurian and the Tyrrhenian Seas (Lisitzin, 1955a). Not less than 10 sea-level stations are in operation in this area and they are comparatively evenly distributed over the entire basin. Exner (1925) developed a method to determine the slope of the water surface. This elementary graphical method may be applied in all cases where there are available sea-level observations from three stations which form the points of a triangle. The vector of the maximum slope of the water surface may then be determined without difficulty. By combining the sea-level stations in a suitable way, 9 different triangles can be constructed for the research area and the vector of the maximum slope within these triangles can be computed for every month. These computations have rendered the following results. In the northern parts of the Tyrrhenian Sea the vector of the slope is approximately 2.5 times greater than in the southern parts of the sea. The reason for this inequality is probably the fact that the sea is narrower in the north than in the south. The seasonal deviations are also very pronounced, showing more-or-less similar features in the whole sea basin. The absolute maximum occurs in August and is particularly pronounced in the northern parts of the sea. The secondary maximum is more dispersed, but occurs chiefly in December and January. Concerning the direction of the greatest slope, the computations have shown the existence of distinct seasonal variations. It can be established that there exists a marked conformity between the months November to May on the one hand, and the period June to October on the other hand. However, the angles of the maximum slope deviate in some cases in May and in October and November from the average of the group concerned, which indicates that the shift of the seasons is not pronounced. In order to demonstrate the above, the average values of the direction of the greatest slope of the water surface in the different parts of the central region of the Tyrrhenian Sea for the cold and warm seasons have been determined, giving the following results ($N = 0°$, $E = 90°$, etc.):

	NE	NW	SE	SW
November – May	28°	98°	23°	99°
June – October	209°	277°	194°	278°
Difference	181°	179°	171°	179°

The difference between the direction for the cold and the warm season is fairly close to 180°. The conditions are thus reciprocal during the two seasons. The angles giving the maximum inclination of the water surface show, in addition, another interesting feature. It may be noted that the maximum slope tends to turn, in stationary cases, perpendicularly to the direction of the coast line. The difference of from 70° to 80° between the western and the eastern parts of the basin may thus be accounted for by the coastal configuration of the sea.

To turn from the values of the most pronounced slope to the determination of the velocity of the current, the first step is to project the vector concerned upon the direction which is perpendicular to the main axis of the basin. In the Tyrrhenian Sea this perpendicular direction is close to the direction N60°E – S60°W. The relationship between the velocity of the current and the perpendicular component of the most marked slope of the water surface has been determined on the basis of eq.1. The average velocity of the current was in the northern parts of the sea basin 1.8 cm/sec and in the southern parts only 0.6 cm/sec. The sign of the computed velocity indicates that the current is normally directed, in the whole sea basin, towards the north during the cold season and towards the south during the months of June to October. In the southern part of the Ligurian Sea the situation is on the whole similar, and the average currents are a continuation of the Tyrrhenian currents. Thus the direction of the currents in the southern part of the sea is orientated towards the west during the winter months and towards the east during the warm season. With respect to the northern part of the Ligurian Sea, the computations have shown that with some less pronounced exceptions the direction of the current is opposite to the characteristic of the southern parts of the sea. This current is sometimes very rapid, the velocity in February reaching for instance as high a value as 4.4 cm/sec.

The above results make it possible to draw an approximate picture of the currents in the Tyrrhenian Sea. Romanovsky (1954) drew the following conclusions on the basis of a very large amount of observed data on currents: 'During summer, a current moves along the coast of northern Africa from Gibraltar to the Strait of Sicily. ... At the height of Bizerta this current divides into two branches, one of them progressing towards the eastern Mediterranean, the other into the Tyrrhenian Sea (0.4 knots). The latter branch turns at the height of Palermo towards the northwest and passes through the Tyrrhenian Sea with a velocity of 0.2 knots. Along the Italian coast runs towards the southeast a counter-current and another proceeds towards the south along the coast of Sardinia. The main current penetrates into the Gulf of Genoa between Corsica and the Isle of Elba, it follows the coast of Italy and thereafter the French Côte d'Azur. The difference between the summer regime and that in winter is not marked'.

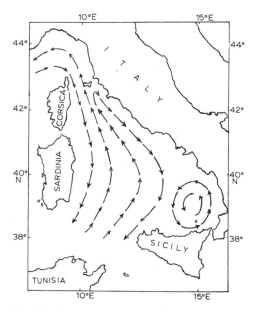

Fig. 19. The average current pattern in the Tyrrhenian Sea and the Ligurian Sea during summer computed from sea-level data (Lisitzin, 1955a).

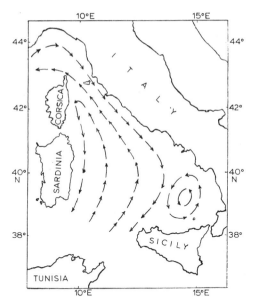

Fig. 20. The average current pattern in the Tyrrhenian Sea and the Ligurian Sea during winter, computed from sea-level data (Lisitzin, 1955a).

It may be of interest to compare the results achieved by Romanovsky with the charts in Fig. 19 and 20, representing the results based on sea-level data. The former of these figures refers to summer, the latter to winter. During the summer months the average current in the Tyrrhenian Sea is directed towards the south. This result does not disagree with the regime outlined by Romanovsky, since the main current which runs northwards is accompanied by the two currents moving along the coasts. It must then be assumed that the sum of the volumes of these two currents is greater than the volume of the current progressing in the middle parts of the sea basin towards the northwest, resulting in a water transport towards the south. If one starts from the average velocity of the current, according to Romanovsky 0.2 knots, corresponding to 10 cm/sec, it is possible to estimate, on the basis of the computed values for current velocity, that in the northern parts of the Tyrrhenian Sea the central current occupies in August about 30% of the breadth of the basin, while its width in the southern parts of the sea increases to 45%. In January the breadth of the central current is, according to the results based on sea-level data, 60% in the north and roughly 52% in the south. It may thus be established that the difference between summer and winter is very pronounced in the northern parts but diminishes considerably towards the south. The most difficult task was the determination of the current direction in the waters between Sicily and the Italian mainland. In this part of the sea basin the velocity of the current varied considerably from one month to another and its direction did not show any greater constancy, which indicates that the stability of the current is slight. The most acceptable scheme was in this area an anticlockwise eddy.

In the Ligurian Sea the current situation changes completely from summer to winter. Overall, the water circulation in the Ligurian Sea is in summer a part of the circulation in the western parts of the Tyrrhenian basin; in winter, on the contrary, it participates in the circulation of the waters in the eastern parts of the basin.

The general scheme outlined in the two charts is, self-evidently, only approximate. According to this scheme, the density of the water should be on average more pronounced along the coast of the Italian mainland and in the area situated between the first and the second current line, counted from the west in Fig. 19 and 20. Although it is always hazardous to compare average conditions with special cases, the profiles given by Aliverti et al. (1968) in the Atlas of the Tyrrhenian Sea seem to confirm the above results.

EVAPORATION AND PRECIPITATION

While the distribution of atmospheric pressure and wind stress affect the relative heights of sea level in a given region, and density variations cause changes in the volume of the water, the processes of evaporation and precipitation, in addition to continental river discharge, bring about fluctuations in the quantity of the water masses. The process of evaporation involves a decrease of the total water in the oceans and seas, while an increase is connected with precipitation and continental run-off. Since the total water masses in the world oceans, if the eustatic increase in water amount is left out of consideration (p. 177–183), remain practically unchanged over a more prolonged period, the three factors mentioned above must be in balance. On the other hand, these factors, although reaching considerable quantities in the course of a period as long as one year, show rather slight variations in particular cases and may, therefore, be studied as average values.

Evaporation is a factor for which the amount is difficult to determine in the open sea by direct measurements. On the other hand, measurements of evaporation made along the coasts of continents and on islands are not representative of the conditions prevailing at sea. Moreover, Wüst (1920) had already pointed out that the amount of evaporated water measured by means of an evaporation vessel on board a ship is, on average, more than twice as large as the actual evaporation at the surface of the sea. This result has been confirmed by more recent observations to be of the correct order of magnitude.

The empirical formula, on which the determination of evaporation E is based, may be written in the form:

$$E = k \, (e_w - e_a) \, W_a \qquad\qquad (1)$$

where k is a factor which for hourly values may be expected to lie between $2.8 \cdot 10^{-6}$ and $9.8 \cdot 10^{-6}$, if E is expressed in g/cm^2 or cm, e_w and e_a represent vapour pressure at the sea surface and at the height a above this surface, in mbar, and W_a the wind velocity at the same height a in cm/sec. The lower value of the coefficient k refers to a smooth surface of the sea, the upper limit to a very rough surface. The value $k = 6 \cdot 10^{-6}$ may thus be applied in average conditions. If in eq. 1 the amount of evaporation in 24 h. is expressed in mm and the wind velocity W_a in m/sec, the equation may be written as:

$$E = 0.143 \, (e_w - e_a) \, W_a \tag{2}$$

On the basis of the above formula, Jacobs (1951) computed the average values of evaporation for the North Atlantic and the North Pacific Oceans and for the two oceans as a total during particular seasons, obtaining the results which are reproduced in Table XVII, converted into mm per day. According to Jacobs the total annual evaporation per unit area is in the North Atlantic Ocean equal to 114.8 cm/year, in the North Pacific Ocean 111.4 cm/year and the average for the two oceans 112.5 cm/year.

The values given above coincide fairly well with the results of Wüst (1954) for the total world oceans and adjacent seas. A digest of Wüst's result is reproduced in Table XVIII, which not only gives the average amounts of evaporation for the particular water-covered zones but also the corresponding values for precipitation. In addition, the table gives the differences between evaporation and precipitation for the latitudinal zones considered.

Table XVIII shows that precipitation has three zonal maxima, corresponding to the latitudes of $50°-40°N$, $10°-0°N$ and $40°-50°S$. The maximum belts in the middle latitudes of the two hemispheres are secondary ones with the precipitation reaching approximately 65% of the quantity characteristic of the main maximum zone situated somewhat to the north of the equator. The zones typical of minimum precipitation are situated between the latitudes $30°-20°N$ and $20°-30°S$. In addition to a relatively narrow belt covering the $10°$ immediately to the north of the equator, the differences between evaporation and precipitation are negative in the higher latitudes to the north and south of the latitudes of $40°N$ and $40°S$ respectively. The total excess of evaporation with respect to precipitation, amounting to slightly more than 10 cm per year, must be compensated by the discharge from rivers.

For particular zones the deviations between evaporation and precipitation are much more pronounced. For the latitudinal belt situated between $30°N$ and $20°N$ evaporation exceeds precipitation by 75 cm per year, while in the zone between $50°S$ and $60°S$ the former term is weaker by 71 cm per year than the latter term. It is highly probable that

TABLE XVII

EVAPORATION (MM/DAY) IN THE NORTH ATLANTIC, THE NORTH PACIFIC AND THE TOTAL IN THE TWO OCEANIC REGIONS DURING DIFFERENT SEASONS

Season	North Atlantic	North Pacific	The two oceans
Winter	4.1	3.6	3.8
Spring	2.8	3.0	2.9
Summer	2.3	2.5	2.4
Autumn	3.2	3.2	3.2
Year	3.1	3.1	3.1

TABLE XVIII

EVAPORATION, PRECIPITATION AND THEIR DIFFERENCES FOR LATITUDINAL ZONES IN
THE TOTAL WORLD OCEANS (IN CM/YEAR)

Zone	Evaporation	Precipitation	Evaporation – precipitation
North			
70° – 60°	16	50	−34
60° – 50°	45	91	−46
50° – 40°	75	107	−32
40° – 30°	116	80	36
30° – 20°	135	60	75
20° – 10°	133	86	47
10° – 0°	119	166	−47
70° – 0°	110.6	101.0	9.6
South			
0° – 10°	131	107	24
10° – 20°	138	85	53
20° – 30°	128	66	62
30° – 40°	103	80	23
40° – 50°	67	114	−47
50° – 60°	25	96	−71
0° – 60°	102.1	91.4	10.7

these differences exert some effect upon the sea level, but the levelling with the adjoining areas will occur fairly rapidly, thus allowing no opportunity to measure this effect.

Returning to the empirical formula, as given in eq. 2, it may be mentioned that the coefficient varies according to different authors within the limits of 0.10 and 0.15. Besides the value of 0.143 given in the formula and used by Jacobs in his computations, several other values may be mentioned. Petterssen et al. (1962) considered 0.130 and Budyko (1956) 0.135 as more probable values. For the Baltic Sea the value of the coefficient concerned was considered to be still lower. Thus Simojoki (1948) based his computations on the coefficient being equal to 0.110, Brogmus (1952) used values between 0.110 and 0.118 and Hankimo (1964) the value 0.114.

There exists, however, another possibility of evaluating the difference between evaporation and precipitation in regions with insufficient direct or indirect measurements of these quantities. Benton and Estoque (1954) pointed out that the difference between evaporation and precipitation per unit area and time, $E - P$, could be determined by using the total water budget in the atmosphere lying above the region under consideration. In this case the following equations are valid:

$$E - P = R_a + S_a \tag{3}$$

$$P - E = R_s + S_s \tag{4}$$

where R_a and R_s denote the run-off from the atmosphere and the sea region, respectively, while S_a and S_s correspond to the changes of the water storage in the atmosphere and sea region. Since R_a may be determined from the outflux of water, primarily in the form of water vapour, from the atmospheric region, and S_a from the change of atmospheric water content, this method makes it possible to estimate the difference $E - P$ for regions with available aerological data. By combining eq. 3 and 4 we obtain:

$$S_s = -(S_a + R_a + R_s) \qquad (5)$$

In this formula the change of the water storage in the atmosphere S_a is generally small owing to the weak water capacity of the atmosphere. Hence the formula may be written approximately:

$$S_s = -(R_a + R_s) \qquad (5a)$$

where, it may be stated again, R_s represents the net outflux of water from the sea region and R_a the net outflux of the atmospheric water in different forms from the corresponding atmospheric region. Fig. 21 gives a schematic picture of the total water balance.

Eq. 5a presents a theoretical possibility for computing the total change of water quantity in a fixed sea region. Since, however, there is hardly any practical possibility of estimating R_s from available observations, it is not possible to evaluate the contribution of the difference between evaporation and precipitation to sea-level changes, even if satisfactory values of this quantity could be reached by using eq. 3. As may be seen from Table XVIII, the quantity $E - P$ shows a highly pronounced latitudinal variation, which, however, must be compensated by water exchange in the oceans during a whole year. Whether the quantity $E - P$ has any appreciable influence upon the sea-level changes during shorter periods of time, for instance during seasons or months, is therefore questionable.

The atmospheric water budget method, as already mentioned above, was originally proposed by Benton and Estoque and has been applied by several scientists for the computation of the quantity $E - P$. For more details reference may be made to Palmén (1967). Starr and Peixoto (1958), and Starr et al. (1965) in particular extended their computations to a global scale showing that the method under certain conditions may give results wholly comparable with those referred to previously. Since, however, the

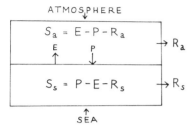

Fig. 21. Schematic picture of the total water balance.

method can be successfully applied only in regions with sufficiently reliable aerological observations, its usefulness is still limited to special regions. Such a region is, for instance, the Baltic Sea. Computations carried out in this region by Palmén and Söderman (1966) showed that the aerological method can result in at least as satisfactory data on the seasonal evaporation from the sea as those computed with the help of empirical formulae. Although this method is highly exact in principle, its usefulness depends on the reliability of the determination of the transport of the water amount in the atmosphere on the basis of data from relatively few aerological stations. The method therefore gives good results for more prolonged time-spans, such as a month or a season, while for shorter periods the errors frequently tend to become more pronounced.

Since the storage and the outflux of water include water not only in gaseous but also in liquid and solid phases, a knowledge of the total water content in the atmosphere is necessary. Aerological data do not give any information on the water quantity in the clouds. This quantity is, as a rule, slight in comparison with the content of water vapour, and may therefore be neglected, especially if the computations cover a relatively large area. The procedure corresponds to an assumption that all condensed water vapour disappears from the atmosphere as precipitation. This assumption is, of course, not valid in all cases. The equation on which Palmén and Söderman based their computations reads:

$$\bar{E} - \bar{P} = \frac{1 \cdot}{g} \int_0^{p_0} \frac{\partial q}{\partial t} \, dp + \frac{1}{gA} \int_0^{p_0} \oint (qv)_n \, dL \, dp$$

In this equation $\partial q/\partial t$ denotes the change in specific humidity, integrated from the top ($p = 0$) to the bottom of the atmospheric column ($p = p_0$), L is the total length of the perimeter of the area A over which the computations are extented and $(qv)_n$ is the flux of water vapour per unit mass normal to the boundary, while g is, as before, the acceleration of gravity.

As mentioned above, the calculations performed by Palmén and Söderman covered the region of the Baltic Sea proper. This area formed a polygon with six aerological stations as points: Stockholm, Copenhagen, Greifswald, Kaliningrad, Riga and Jokioinen. The relatively large land area (22.8% of the total area) covered by the polygon had evaporation rates deviating from those on sea and so required the application of a special correction in order to determine the amount of evaporation from the sea surface.

According to the equation given above, the computations allow the determination of the difference between evaporation and precipitation. Since precipitation is known, as a rule, for 12-h periods, the amount of evaporation E of the difference $E - P$ refers also to periods covering 12 h.

Precipitation was determined as the arithmetical average of the 27 precipitation stations operating in the region. Several of these stations are located on islands in the central parts of the Baltic proper, giving thus a concept of the amount of the average precipitation over the sea surface.

In Table XIX is given a comparison between the monthly values of precipitation

TABLE XIX

MONTHLY AVERAGE PRECIPITATION (MM) IN THE AREA *A* FOR THE PERIOD OCTOBER 1, 1961, TO SEPTEMBER 30, 1962, COMPARED WITH THE CORRESPONDING PRECIPITATION FOR THE 50-YEAR PERIOD 1886 – 1935

	J	*F*	*M*	*A*	*M*	*J*	*J*	*A*	*S*	*O*	*N*	*D*	*Year*
Oct.1, 1961 – Sept.30, 1962	55	36	20	35	55	38	51	84	68	40	44	43	569
1886–1935	46	32	32	37	35	38	55	66	52	57	52	52	554
Difference	9	4	–12	–2	20	0	–4	18	16	–17	–8	–9	15

determined by Palmén and Söderman for the period extending from October 1, 1961, to September 30, 1962 and used by them in their calculations, and the average monthly precipitation over the Baltic for the 50-year period 1886–1935 as computed by Simojoki (1949). Although no complete correspondence could be expected, the agreement is on the whole quite satisfactory. Especially remarkable is the close coincidence between the two values which give the annual quantity of precipitation.

Before the year 1966, when the computations by Palmén and Söderman were performed and published, all determinations of evaporation in the Baltic Sea were made with the help of empirical formulae of the type given in eq. 1. Since reliable results may be achieved on the basis of these formulae, it may be of interest to compare the results of Palmén and Söderman with previous determinations.

Simojoki (1948) computed the average monthly values of evaporation for the Bogskär lighthouse, situated in the northern part of the Baltic proper, for a period covering 13 years. Brogmus (1952) performed similar computations for the central parts of the Baltic Sea. In Table XX are compiled the results obtained by Simojoki and Brogmus for comparison with the values of evaporation determined by Palmén and Söderman.

The coincidence between the particular values in Table XX is surprisingly good, particularly if it is taken into consideration that the values computed by Palmén and Söderman refer to one year only. The authors themselves pointed out that the relatively low value of evaporation in October 1961 and the pronounced evaporation in December of the same year may be easily accounted for by the quite special meteorological conditions which were characteristic of these months. In October 1961 the air temperature was unusually high suppressing evaporation, while in December several strong fronts of cold air were observed, water temperature still being high as a consequence of the preceding warm autumn months.

It may appear that too much space has been dedicated here to the problem of evaporation and precipitation, since their share and significance in sea-level variations as a whole are not very pronounced. Nevertheless, it must always be kept in mind that the question of the water balance in a semi-enclosed sea basin like the Baltic Sea is developing into one of the most important problems, for instance, in connection with the study of pollution in seas of this type. This problem may to some extent be satisfactory solved on the basis

TABLE XX

THE MONTHLY EVAPORATION (MM) FROM THE BALTIC SEA ACCORDING TO DIFFERENT
AUTHORS

	Empirical equations		Water vapour transport
	Simojoki (1948)	Brogmus (1952)	Palmén and Söderman (1966)
January	52	45	67
February	46	36	45
March	35	19	40
April	16	10	22
May	6	15	6
June	6	24	16
July	24	40	35
August	67	55	45
September	50	55	43
October	58	81	44
November	77	74	74
December	75	60	91
Year	512	514	528

of adequate sea-level data, but evaporation and precipitation become significant factors,
as soon as the contribution of the river discharge and the amount of water inflow and
outflow in a semi-enclosed sea region have to be taken into account.

CHAPTER 4

SEASONAL VARIATIONS

THE SEASONAL CYCLE IN SEA LEVEL

The question of the seasonal cycle in sea level in oceans and seas has been studied by a considerable number of different oceanographers. One of the first papers concerned with this problem was that published by Nomitsu and Okamoto (1927). In their study on the causes of the annual variations of mean sea level along the Japanese coasts, the two authors made a first attempt to determine the contribution of different factors to these variations. In their efforts to compute the relationship between wind and sea level, Nomitsu and Okamoto based the results on the formula given by Colding (1880):

$$H = 7.63 \cdot 10^7 \frac{L}{d} W^2 \sin^2 \alpha$$

where H is the difference in sea level in metres, caused by the piling-up effect of the wind at opposite coasts with reciprocal distance amounting to L, d is the average depth of the sea, W the velocity of the wind, and α the angle between the direction of the wind and the coastline. The computations showed that the long-term seasonal departures from the mean sea level along the Japanese coasts were as follows:

Winter	Spring	Summer	Autumn
1.3 cm	−0.5 cm	−0.9 cm	0.2 cm

These departures are not very pronounced. The amplitude of the variation is considerably less than the recorded one, and, above all, the sign of the departures is opposite to that of the observations. These results indicate that the contribution of the wind is not significant. On the other hand, Nomitsu and Okamoto were able to show that the static effect of atmospheric pressure corresponds approximately to 20% of the observed amplitudes of mean sea-level variations. Finally, Nomitsu and Okamoto drew the conclusion that the fluctuations in water density account for more than 80% of the sea-level variations in the Sea of Japan. La Fond (1939) was also able to show that at La Jolla (California) the changes in sea level may be attributed entirely to the variations in water density.

Quite independently of the results described above, studies based on sea-level data recorded along the European coasts have shown that the seasonal variations are characterized by a fairly pronounced regularity. It can be proved that the maximum level is observed in European waters during the later part of the year and that there is a retardation of the occurrence of the maximum when progressing from the south towards the north. This fact posed the question of whether the seasonal cycle in sea level in other oceans and seas shows a similar tendency. In a study on the seasonal variations in sea level in the Atlantic Ocean, Polli (1942) was able to establish that the maximum sea level along the coasts of Argentina is observed in March, i.e., practically at the same time as the lowest sea level is noted in the northern parts of the Atlantic Ocean and adjacent seas.

A more extensive study of the problem on a world-wide scale would assuredly involve considerable difficulties. Sea-level data are fairly numerous and reliable for some parts of the oceanic coasts, but they are almost completely lacking for other parts of the oceans, or they are not easily accessible and frequently not very accurate. Useful assets in this connection are the publications on mean sea level regularly issued by the Assocation of Physical Oceanography (Association d'Océanographie Physique, 1940, 1950, 1953, 1958, 1959, 1963, 1968). These publications give the monthly and annual mean sea-level heights for a great number of stations. In spite of the fact that for some of the stations the data are rather scarce, covering only one year or a few years, a critical use of the data makes it possible to obtain a more-or-less satisfactory picture of the general features characteristic of the particular parts of the oceans and seas. A certain disadvantage concerning the data mentioned above is the lack of simultaneous observations. This is particularly true for the data covering the time before the year 1940 where the periods may in some cases differ very markedly from each other. In addition, in all the cases where the time covered by the data is restricted the uncertainy of the computed results may be rather pronounced and there is always the risk that the application of some levelling method, for instance, of harmonic analysis will give a distorted picture of the phenomenon as a whole. On the other hand, it must always be kept in mind that if the results agree relatively well for different periods, this implies that the general features of the seasonal cycle in sea level are more or less stable.

A few typical examples may be sufficient to illustrate the range of possible deviations arising from the use of different periods in cases where the observed data may be considered fairly reliable. For the sea-level stations situated on the coasts of the Atlantic Ocean, data from Brest in France are at our disposal. The observations started as early as 1807 and continued without interruption until 1835. Moreover, the data from 1846 to 1856 and from 1861 to 1943 were available and analyzed. The analyzed data cover altogether a time-span of 123 years. Basing the results on the entire series, the harmonic analysis of the annual cycle gives November 13 as the date of the maximum sea level. Taking into account only the observations obtained before 1891, this date occurs a week earlier i.e., November 6, while for the more recent period extending over the years 1891—1943 the maximum sea level occurs, according to the results of the harmonic analysis, on November 22. Two completely different periods, covering 70 and 53 years respectively, thus result in a time deviation amounting to 16 days. The scattering becomes much more pronounced if particular years are considered especially those representing the extreme cases. The harmonic analysis of the data for the year 1826 indicates September 13 as the date of the occurrence of the maximum sea level, while for the year 1879 the corresponding result is April 17. For the remaining years the date of the maximum was noted in autumn, winter or spring. On the contrary, the summer months never seem to produce the highest monthly mean sea level of the year.

In addition to the departures brought about by different disturbing factors, which show a more-or-less pronounced variability during particular years, the sea level may be influenced by the local conditions characteristic of the station. One station may, for instance, be exposed very markedly to the effect of the predominant wind, while this effect is not too significant for some other station in the neighbourhood. Without possessing knowledge of the positions of the different sea-level stations, the data of which have been utilized for the determination of seasonal variations, it is completely beyond the limits of possibility to make the corrections which may be necessary.

All the contributing factors mentioned above result in the fact that the harmonic constants of the annual cycle in sea level, determined for the different parts of the oceanic coasts, may present only an approximate picture of the actual conditions (Lisitzin, 1955c). A highly comprehensive and critical investigation is indispensable for this purpose. It is also necessary to leave out of consideration a number of the harmonic constants, which are based on as short a period as one year, or at the utmost a few years. It did not seem appropriate therefore to include in the following tables a number of values of the annual constituent which have been published by the International Hydrographic Bureau in Monaco. However, it has been possible to use important information given by the Bureau for regions with practically no available sea-level data for the investigation.

The harmonic constants determind for the annual cycle in sea level are reproduced in Tables XXI–XXIII. These tables refer to the Atlantic Ocean, the Pacific Ocean and the Indian Ocean respectively, also covering in each case a number of adjacent seas. The harmonic constants for stations marked with an asterisk have been taken directly from the tables published by the International Hydrographic Bureau. In addition to the names of stations, an indication of the number of years on which the determination of the harmonic constants is based, the geographical co-ordinates of the stations, the amplitudes, the phases and the dates corresponding to the maximum height of the seasonal cycle are given in the tables.

Studying the values for the amplitudes more closely, one may easily note that they are characterized by a considerable dispersion, probably brought about by the effect of different local factors. These values are therefore, as a rule, not adequate to give more general information. Conversely, the phases and the corresponding dates for the occurrence of the maximum sea level, in spite of considerable deviations, present interesting and important results.

The Atlantic Ocean

Table XXI starts with the stations Teriberka and Liinahamari, which are situated on the coasts of the Arctic Ocean. Since this ocean forms an immediate continuation of the northernmost part of the Atlantic Ocean, it may easily be supposed that the two regions

TABLE XXI

THE HARMONIC CONSTANTS OF THE SEASONAL VARIATION IN SEA LEVEL IN THE ATLANTIC OCEAN

Station	Number of years	Latitude	Longitude	Amplitude in cm	Phase in degrees	Date of maximum
Teriberka Bay *	1	69°11'N	35°08'E	8.8	242	Nov. 23
Liinahamari	8	69°39'N	31°22'E	11.2	235	Nov. 16
Aberdeen	52	57°09'N	02°05'W	9.0	241	Nov. 22
Liverpool *	7	53°25'N	03°00'W	12.0	229	Nov. 10
Dublin	9	53°21'N	06°13'W	6.5	234	Nov. 15
Den Helder *	20	52°58'N	04°45'E	9	224	Nov. 5
IJmuiden *	20	52°28'N	04°34'E	11	254	Dec. 5
Hook of Holland *	20	51°59'N	04°07'E	8	254	Dec. 5
Newlyn	31	50°06'N	05°33'W	4.2	233	Nov. 14
Brest	123	48°23'N	04°29'W	4.9	232	Nov. 13
Cascais	30	38°41'N	09°25'W	3.1	193	Oct. 4
Lagos	30	37°06'N	08°40'W	5.0	188	Sept. 29
Horta	27	38°32'N	28°38'W	4.0	185	Sept. 26
Angra do Heroismo	4	38°38'N	27°12'W	4.4	207	Oct. 18
Santa Cruz	8	28°29'N	16°14'W	6.2	155	Aug. 26
Freetown *	1	08°30'N	13°14'W	5.2	164	Sept. 5
Takoradi	17	04°53'N	01°45'W	6.4	296	Jan. 17
Accra	1	05°32'N	00°12'W	7.7	295	Jan. 16
Cape Town *	1	33°54'S	18°25'E	3.4	76	June 7
Simons Bay *	1	34°12'S	18°26'E	3.7	355	March 16
Comodoro Rivadavia	17	45°52'S	67°29'W	3.4	321	Feb. 12
Belgrano	20	38°53'S	62°06'W	7.5	335	Feb. 26
La Plata	19	34°55'S	57°56'W	7.3	325	Feb. 16
Buenos Aires	41	34°36'S	58°22'W	7.6	314	Feb. 5
Puerto de Colonia	9	34°29'S	57°51'W	6.6	323	Feb. 14
Santos*	1	23°56'S	46°19'W	7.4	25	April 16
Cabedello *	1	06°58'S	34°50'W	1.1	32	April 23
Belem *	1	01°27'S	48°30'W	14.8	357	March 18
Port of Spain	1	10°39'N	61°31'W	12.0	167	Sept. 8
Zapara Island	6	10°58'N	71°34'W	7.3	181	Sept. 22
Galveston	38	29°19'N	94°47'W	6.9	152	Aug. 23
Key West	20	24°33'N	81°48'W	8.6	185	Sept. 26
Miami Beach	15	25°46'N	80°08'W	8.8	199	Oct. 10
Fernandina	26	30°41'N	81°28'W	9.0	184	Sept. 25
Ireland Island	10	32°19'N	64°50'W	7.5	215	Oct. 26
St. Georges	3	32°22'N	63°42'W	6.0	194	Oct. 5
Charleston	25	32°47'N	79°55'W	8.5	174	Sept. 15
Hampton Roads	17	36°57'N	76°20'W	6.1	177	Sept. 18
Baltimore	44	39°16'N	76°35'W	12.4	129	July 31
Atlantic City	33	39°21'N	74°25'W	7.3	147	Aug. 18
Port Hamilton	28	40°37'N	74°02'W	8.9	137	Aug. 8
Boston	25	42°21'N	71°03'W	3.3	138	Aug. 9
Portland	34	43°40'N	70°15'W	3.4	129	July 31
Eastport	17	44°54'N	66°59'W	0.5	145	Aug. 16
St. John *	20	45°15'N	66°04'W	4.0	125	July 26

TABLE XXI (continued)

Station	Number of years	Latitude	Longitude	Amplitude in cm	Phase in degrees	Date of maximum
Halifax	13	44°40'N	63°34'W	4.3	276	Dec. 27
St. Paul Island	5	47°12'N	60°09'W	3.4	247	Nov. 28
Port aux Basques	2	47°34'N	59°07'W	1.5	262	Dec. 13
Harrington	5	50°31'N	59°27'W	4.7	241	Nov. 22

show features which are more or less identical. The harmonic constants in Table XXI prove this fact very distinctly. It may be noted that the phases of the two stations are not only in keeping with each other, but also with those of the eight stations situated to the north of the latitude 40°N on the east coast of the Atlantic Ocean and adjacent seas. On average, the date of the maximum sea-level height for the whole region is November 19, and the largest deviation amounts to 30 days.

Following the eastern coast of the Atlantic towards the south, it may be established that the six stations from Cascais to Freetown form, as far as the phases are concerned, a special group, which although showing considerable deviations for particular stations, differs distinctly from the first group. In this connection it must be pointed out that this second group does not extend as far as the equator, but approximately to the latitude of 6°N. The average data for the maximum sea level for this group is September 23.

If the difference is already pronounced between the first and the second group, it becomes still more accentuated when taking into consideration the stations situated on the west coast of Africa between the latitudes of 6°N and 35°S. Admittedly, it covers an extensive region which is only weakly represented. Not less than three of the four stations have results based on observations made during one year only. In particular the date for Cape Town seems not to be very reliable and it has therefore not been considered when determining the average date for the group. The three remaining stations give February 6 as a result.

Considering next the stations situated on the American coast of the Atlantic Ocean, it can be noted immediately that they are in almost complete conformity with the general features described above. The southernmost part of this region is well represented by the stations in Argentina while the two stations Santos and Cabedello are probably less reliable. These two stations are responsible for the considerable dispersion of the dates of the maximum sea level, corresponding to not less than 78 days. The average date, being March 5 for the whole region is probably somewhat retarded.

Between the latitudes of 10°N and 45°N, there are in Table XXI not less than 17 stations with sea-level data based on a satisfactory number of years. In spite of this fact the dispersion is fairly pronounced along this coastal region, reaching in some cases 90 days. Nevertheless, there is a marked tendency indicating a retardation of the date of the maximum when progressing from the north towards the south. The general trend is thus opposite to that noted along the European coasts. Moreover, there seems to be an indication that for islands situated at some distance from the coast, i.e., Ireland Island

TABLE XXII

THE HARMONIC CONSTANTS OF THE SEASONAL VARIATION IN SEA LEVEL IN THE PACIFIC
OCEAN

Station	Number of years	Latitude	Longitude	Amplitude in cm	Phase in degrees	Date of maximum	
Seeward	11	60°06'N	149°27'W	11,2	253	Dec.	4
Ketchikan	28	55°20'N	131°38'W	10.4	272	Dec.	23
Port Simpson *	10	54°34'N	130°26'W	12.8	270	Dec.	21
Prince Rupert *	10	54°19'N	130°19'W	11.3	268	Dec.	19
Wadhams *	5	51°24'N	127°38'W	12.5	280	Jan.	1
Clayoquot *	11	49°09'N	125°55'W	14.3	280	Jan.	1
Vancouver *	18	49°18'N	123°07'W	6.7	260	Dec.	11
Sand Heads *	12	49°07'N	123°18'W	6.3	253	Dec.	4
Victoria *	15	48°26'N	123°23'W	9.4	276	Dec.	27
Seattle	48	47°36'N	122°20'W	7.1	293	Dec.	14
Astoria Town *	2	46°11'N	123°50'W	7.4	284	Dec.	5
San Francisco	49	37°48'N	122°28'W	3.3	222	Nov.	3
Los Angeles	23	33°43'N	118°16'W	6.8	185	Sept.	26
San Diego	20	32°42'N	117°14'W	6.5	186	Sept.	27
Mazatlan *	1	23°11'N	106°27'W	3.8	153	Aug.	24
Honolulu	42	21°18'N	157°52'W	3.4	194	Oct.	5
Balboa	38	08°58'N	79°34'W	12.6	169	Sept.	10
Valparaiso*	1	33°02'S	71°38'W	4.6	351	March	12
Orange Bay *	1	53°31'S	68°05'W	4.8	92	June	23
Bluff *	5	46°36'S	168°20'E	2.7	17	April	8
Dunedin	5	45°53'S	170°33'E	1.5	4	March	25
Port Lyttelton	5	43°36'S	172°51'E	5.0	5	March	26
Wellington	10	41°17'S	174°48'E	1.5	9	March	30
Auckland *	7	36°51'S	174°46'E	5.7	31	April	22
Ballina	5	28°52'S	153°37'E	7.1	45	May	6
Sydney	31	33°51'S	151°14'E	3.3	28	April	19
Port Adelaide	10	34°51'S	138°30'E	6.0	103	July	4
Thursday Island *	1	10°35'S	142°13'E	15.4	315	Feb.	6
Manado	5	01°32'N	124°50'E	2.9	358	March	19
Port Uson	3	12°01'N	120°12'E	7.8	180	Sept.	21
Manila	25	14°35'N	120°58'E	13.0	145	Aug.	16
Macao	10	22°12'N	113°33'E	8.4	207	Oct.	18
Takao	30	22°37'N	120°16'E	9.8	151	Aug.	22
Horosima	30	32°25'N	131°40'E	14.3	151	Aug.	22
Hukabori *	5	32°41'N	129°49'E	17.0	149	Aug.	20
Kusimoto	25	33°28'N	136°47'E	13.7	155	Aug.	26
Kobe *	6	34°41'N	135°11'E	17.0	149	Aug.	20
Tonoura *	6	34°55'N	132°04'E	18.5	153	Aug.	24
Aburatubo *	5	35°10'N	139°37'E	9.2	175	Aug.	16
Yokohama *	6	35°27'N	139°38'E	10.5	168	Aug.	9
Wazima *	5	37°24'N	136°51'E	16.6	165	Aug.	6
Aikawa	25	38°18'N	141°30'E	9.9	180	Aug.	21

TABLE XXII (continued)

Station	Number of years	Latitude	Longitude	Amplitude in cm	Phase in degrees	Date of maximum	
Iwasaki *	6	40°35'N	139°54'E	12.8	168	Aug.	9
Ominato Ko *	5	41°15'N	141°08'E	11.1	168	Aug.	9
Hanasaki	25	43°17'N	145°35'E	3.6	228	Nov.	9
Otomari Ko *	1	46°38'N	142°45'E	3.4	220	Nov.	1

and St Georges, the date of the maximum occurs later than for the continental stations. The average date for the maximum for the entire region is September 4.

For the northernmost region on the American coast of the Atlantic Ocean, represented by four stations, it may be noted that there is a considerable correspondence with the stations of the northernmost group on the European coast, in spite of the fact that the maximum occurs on average 18 days later, i.e., December 7. In this connection it must be pointed out that the station Halifax belongs distinctly to the last-mentioned group, although it is situated more to the south than, for instance, the two stations Eastport and St. John which seem to agree better with the previous group. The explanation of this phenomenon must probably be sought in the fact that the position of Halifax is more exposed to outer disturbing effects, while the two latter stations have a relatively well-protected position in the Bay of Fundy.

The Pacific Ocean

The results outlined above obtained for the Atlantic Ocean may be repeated almost completely for the Pacific Ocean. Table XXII shows the occurrence of a region on the American coast of the ocean where the maximum sea level is observed in December, but in some exceptional cases on January 1. The average date is December 23. This region extends approximately to the latitude 40°N. The station San Francisco seems to be situated in a transition zone between this region and the next which covers 5 stations and has as the average date of maximum September 18. For this region it may also be established that the date of the maximum is noted in Honolulu, on the Hawaiian Islands, later than at the continental coastal stations. The third region is represented by a sole station, Valparaiso, with sea-level observations covering only one year. The date of the maximum is March 12, but this result must be considered with some reservation. The next station, Orange Bay, with harmonic constants based also on one year only, is very interesting since it represents the southernmost point considered in the study. If the date of the maximum of this station, June 23, is correct, the conclusion could be drawn that there exists a pronounced similarity between the two hemispheres. They should in this case be composed, as far as the seasonal sea-level changes are concerned, of two parts separated approximately along the latitudes of 40°N and 40°S. In the parts characterized by the higher latitudes in each hemisphere, the date of the maximum of the sea level is

TABLE XXIII

THE HARMONIC CONSTANTS OF THE SEASONAL VARIATION IN SEA LEVEL IN THE INDIAN OCEAN

Station	Number of years	Latitude	Longitude	Amplitude in cm	Phase in degrees	Date of maximum
Franklyn Harbour	5	35°41'S	136°56'E	4.6	75	June 6
Thevenard	6	32°09'S	133°39'E	8.0	93	June 24
Samarang	7	07°00'S	110°24'E	3.6	78	June 9
Pekalongan	6	06°52'S	109°40'E	4.8	73	June 4
Tandjonk-Priok	7	06°06'S	106°54'E	4.8	110	July 11
Makassar	7	05°06'S	119°24'E	10.3	144	Aug. 15
Benkoelen	7	03°47'S	102°24'E	2.4	156	Aug. 27
Emmahaven	7	00°58'S	100°20'E	4.3	148	Aug. 19
Belawan	7	03°55'N	98°43'E	9.0	134	Aug. 5
Ko Ta-phao Noi	6	07°50'N	98°26'E	8.7	145	Aug. 16
Port Blair *	41	11°41'N	92°46'E	6.1	151	Aug. 22
Mergui *	5	12°26'N	98°36'E	18.0	146	Aug. 17
Diamond Island *	5	15°52'N	94°17'E	19.3	158	Aug. 29
Amherst *	6	16°05'N	97°34'E	23.1	136	Aug. 7
Moulmein *	12	16°29'N	97°37'E	74.3	149	Aug. 20
Elephant Point *	5	16°29'N	96°18'E	25.7	140	Aug. 11
Rangoon *	41	16°46'N	96°10'E	41.8	151	Aug. 22
Akyab *	5	20°08'N	92°54'E	28.8	145	Aug. 16
Chittagong *	5	22°20'N	91°50'E	47.8	134	Aug. 5
Calcutta *	40	22°32'N	88°20'E	83.7	157	Aug. 28
Dublat Sangor *	5	21°38'N	88°08'E	26.7	151	Aug. 22
Vizagapatan	10	17°41'N	83°17'E	16.7	189	Sept. 30
Madras *	26	13°06'N	80°18'E	11.2	211	Oct. 22
Negapatan *	5	10°46'N	79°51'E	13.5	234	Nov. 15
Tricomalee *	6	08°33'N	81°13'E	7.5	268	Dec. 19
Colombo *	6	06°57'N	79°51'E	9.5	308	Jan. 29
Galle *	6	06°02'N	80°13'E	10.7	309	Jan. 30
Minicoy *	5	08°17'N	73°03'E	11.0	355	March 16
Tuticorin *	5	08°48'N	78°09'E	9.1	310	Feb. 1
Cochin *	6	09°58'N	76°15'E	10.3	302	Jan. 23
Beypore *	6	11°10'N	75°48'E	9.4	311	Feb. 2
Karwar *	5	14°48'N	74°04'E	10.7	310	Feb. 1
Murmagao *	5	15°25'N	73°48'E	8.2	313	Feb. 4
Bombay (Apollo Bondar)	69	18°55'N	72°50'E	2.1	341	March 2
Bhaonagar	10	21°48'N	72°09'E	28.4	160	Sept. 1
Okha Point *	2	22°28'N	69°05'E	4.3	84	June 15
Karachi *	54	24°48'N	66°58'E	4.5	73	June 4
Aden *	42	12°47'N	44°59'E	11.2	346	March 7
Kilindini	1	04°04'S	39°39'E	3.2	2	March 23
Dar-es-Salaam *	1	06°49'S	39°19'E	3.5	311	Feb. 2
Tamatave *	1	18°09'S	49°26'E	5.5	304	Jan. 25
Beira *	1	19°50'S	34°50'E	13.7	329	Feb. 20
Port Louis *	1	20°09'S	57°29'E	6.4	346	March 7
Lourenço Marques *	1	25°58'S	32°34'E	8.3	298	March 19
Durban *	1	29°52'S	31°03'E	7.3	305	March 26

observed two to three months later than in the regions corresponding to the lower latitudes. This hypothesis is strengthened by the fact that for the station Fort Adelaide, located on the southern coast of Australia, the date of the maximum is July 4. The value of the phase for Port Adelaide deviates so markedly from the results for the neighbouring stations, that it may be noted immediately that this station forms a separate group. The remaining nine stations situated along the southwestern coast of the Pacific Ocean show a considerable conformity and result in April 1 as the average date of the maximum. Also in connection with this region it may be noted that it extends a few degrees to the north of the equator, since the station Manado is distinctly a part of it. Progressing farther to the north there is a group consisting of not less than 15 stations and showing a dispersion of the dates of the maximum of 62 days. This pronounced dispersion is, however, due to one station only, Macao. Leaving this station out of consideration the dispersion diminished to 35 days. The average date for the maximum is in this case September 5, while, finally, for the two stations situated to the north of this region the maximum is on average noted as November 5.

The Indian Ocean

Table XXIII refers to the Indian Ocean. The shape and extension of the basin differs considerably from those of the two oceans discussed above. It is, therefore, self-evident that some of the regions described for those oceans are completely lacking in the Indian Ocean. In other respects also the Indian Ocean shows marked deviations from the general features characteristic of the Atlantic and Pacific Oceans. A closer examination of Table XXIII reveals these deviations very distinctly.

The above results allow us to draw some significant conclusions concerning the average dates of the maximum of the sea level in different parts of the Atlantic and Pacific Oceans. In the following compilation are reproduced the average dates of the maximum for the different regions and the average dates for each zone situated between the same latitudes:

Zone	Atlantic Ocean		Pacific Ocean		Average
	E. coast	W. coast	E. coast	W. coast	
ʼ— 40°N	Nov. 19	Dec. 7	Dec. 23	Nov. 5	Nov. 29
40°N — 6°N	Sept. 23	Sept. 7	Sept. 18	Sept. 5	Sept. 13
6°N — 40°S	Feb. 6	March 5	March 12	April 1	March 6
40°S —	—	—	June 23	July 4	June 28

On studying the average values more closely, it may be noted that the dates of the maximum occur at intervals of approximately three months. If levelling slightly these average values, the following results are obtained:

$-40°N$	December 11	$6°N - 40°S$	March 11
$40°N - 6°N$	September 11	$40°S -$	June 11

The greatest difference in comparison with the column "Average" amounts to 17 days and concerns the southernmost zone for which the results are the least reliable, being based on a very restricted number of observations.

On the basis of the above values, and assuming that there are none or only slight changes in the water volume, an approximate scheme for the variation in sea level in the Atlantic and Pacific Oceans may be given:

Zone	11 March	11 June	11 September	11 December
$-40°N$	$-$	min	$+$	max
$40°N - 6°N$	min	$+$	max	$-$
$6°N - 40°S$	max	$-$	min	$+$
$40°S -$	$+$	max	$-$	min

Towards the middle of March occurs the minimum of the sea level in the second zone and the maximum in the third zone. In the first zone the 'minus' sign indicates that the sea level is decreasing, while the 'plus' sign in the fourth zone indicates the opposite. After March 11 the sea level begins to increase in the second zone and to decrease in the third zone. This is at least partly caused by the flow of water (indicated by the arrows) which is directed towards the south and moving from the first zone into the second and from the third zone into the fourth. This flow is a reality in so far that it can be proved by variations in sea level which are the consequence of changes in atmospheric pressure over the oceans from March to June (Lisitzin, 1960). This situation lasts on average until June 11 when the minimum of sea level is reached in the first zone and the maximum in the fourth zone. Starting from this moment the water flow, owing to variations in the distribution of atmospheric pressure over the oceans, changes direction towards the north and reaches especially high values between the third and the second zone, a fact which in the above scheme is indicated by the double arrows. Towards September 11 a situation is reached in which the two hemispheres present the opposite of the conditions characteristic of the northern spring. After this date the water flow towards the north continues, but it is mainly directed from the second zone to the first, and from the fourth zone to the third. After having reached the maximum sea level in the first zone and the minimum in the fourth zone about December 11, a mirror picture has arisen of the features typical of the middle of June. Now the water flow changes its direction once more and the movement is everywhere directed towards the south. Variation in the atmospheric pressure is again the factor which brings about this flow. The strength of the flow is particularly marked during the time-span up to the middle of March between the second and the third zone. In this connection it must once more be pointed out that the above scheme is only approximate, although it seems to reveal the principal features of the seasonal changes in sea level which are the consequences of the variations of atmospheric pressure over the oceans.

However, it must always be remembered that the water flow is only one of the numerous factors affecting sea level and that the concerned changes depend to a high degree, as Nomitsu and Okamoto have shown, upon the fluctuations of the density of sea water. In fact, Pattulo et al. (1955) dedicated an extensive paper to the problem of the contribution of the water density changes to the seasonal oscillation of the sea level. They introduced the term 'steric' sea level. The steric sea level is defined in terms of the seasonal fluctuations in specific volume of the sea water. A comparison between the recorded and the steric sea levels has shown that the departures agree remarkably well in the low and temperate latitudes. In these regions the steric sea-level heights are largely associated with the fluctuations of water temperature in the upper 100-m layer. Conversely, conditions were less easy to distinghuish in the high latitudes. With the exception of pronounced semi-annual variations which were found along the west coast of South Africa, in Indonesia, in the regions in the vicinity of the Labrador and Oyashio currents, in the Gulf of Mexico and at adjoining Gulf Stream stations, the oscillations were largely annual in character with low sea level in each hemisphere during spring and high sea level in autumn. These results are in complete confirmity with the above-outlined features of the seasonal variation in sea level given by Lisitzin. Recorded amplitudes vary, according to Pattullo et al., from a few centimetres in the tropics to a few decimetres in higher latitudes: they exceed one metre in the Bay of Bengal. According to the results obtained in the paper the effect of atmospheric pressure and of long-period astronomic tides accounts for only a slight part of the recorded fluctuations.

In numerous regions the recorded sea-level departures approximately equal the sum of the steric departures and the hydrostatic effect of atmospheric pressure. In such regions the total weight of water and air per unit area remains more or less constant throughout the whole year, and a pressure recorder at the sea bottom would show only small seasonal fluctuations. In analogy with geodetic usage the term 'isostatic' has been introduced by Pattullo et al. to characterize these conditions.

The problem of the seasonal oscillation in sea level for the Pacific Ocean was investigated by Lisitzin and Pattullo (1961) on the basis of the extensive observations performed during the International Geophysical Year (IGY), 1957—1958. The results are mainly based on the recorded sea-level data and compared with the combined effect of the specific volume of sea water and atmospheric pressure. The results are reproduced in Fig. 22—25. The combined effect of the variations in specific volume and atmospheric pressure is given in the charts of these figures by contours. These contours have *not* been drawn to conform with the vertical numbers entered on the charts; these numbers refer to the observed departures in sea level.

The zero lines in the charts are typical of the main features of the seasonal fluctuations. In the charts representing most closely March and September, the zero line seems to be principally latitudinal, but in June and December an additional longitudinal term may be noted. Comparing the charts for March and September with each other and leaving out of consideration the more or less isolated, closed curves in the central parts of the ocean, it may be noted that the general features of the two months are the mirror

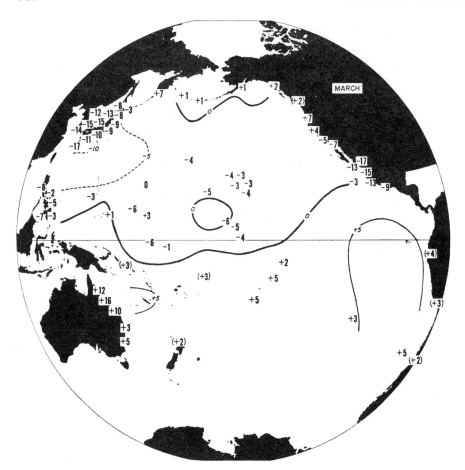

Fig. 22. Curves representing the isostatic deviation from the annual mean sea level for the months February, March and April. Heavy solid lines indicate zero deviation, light solid lines positive deviations and dashed lines negative deviations (in cm). Each curve is labelled in inclined lettering. Vertical numbers show recorded deviations from the annual mean sea level for the same months during the International Geophysical Year. (Lisitzin and Pattullo, 1961; copyright by American Geophysical Union.)

image of one another. In the regions characterized by positive departures in March, there are negative deviations in September and vice versa. Moreover, the magnitude of the departures is similar. An inversion of the deviations may also be observed for the months June and December, although there are more accentuated differences between the shapes of the corresponding curves for these months than for the former pair of months.

Most of the sea-level records given in Fig. 22–25 refer, as already mentioned above, to the International Geophysical Year. They were collected and published by the Permanent Service for Mean Sea Level (Association d'Océanographie Physique, 1959). In order to avoid extensive regions with no sea-level data, monthly mean values for earlier years were entered in the charts for Vancouver, Talara, Matarami, Valparaiso, Samoa, Guadalcanal

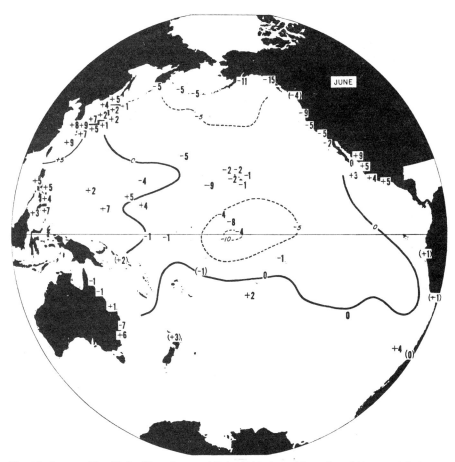

Fig. 23. Same as Fig. 22 for May, June and July. (Lisitzin and Pattullo, 1961; copyright by American Geophysical Union.)

and Auckland (from Pattullo et al., 1955). These data, of non-IGY origin, are given on the charts in parenthesis. Before proceeding to study the figures more closely, it may be pointed out that some of the discrepancies noted in the following are due to the fact that the data are not always synoptic.

The vertical numbers in the charts give the height of the recorded sea level referred to the yearly average value and corresponding to the mean value for the three-month periods February through April, May through July, etc. They represent thus most closely the months March, June, etc.

(1) Fig. 22. February, *March*, April. The main characteristic of this chart is that the isostatic term is negative over most of the North Pacific Ocean and positive in the South Pacific. The most pronounced departures of about −10 cm are noted in the vicinity of Japan. The recorded sea-level heights agree, in general, remarkably well with the isostatic data except along certain coastal lines. The most marked deviations occur along the coast

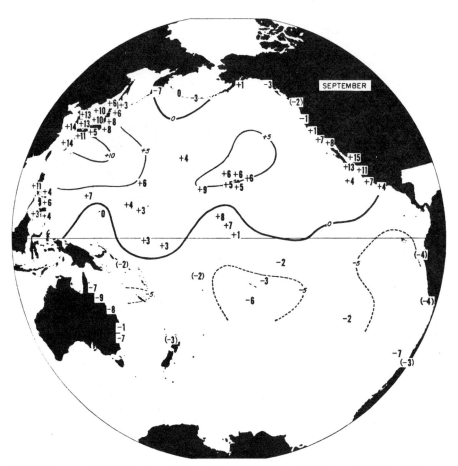

Fig. 24. Same as Fig. 22 for August, September and October. (Lisitzin and Pattullo, 1961; copyright by American Geophysical Union.)

of Mexico, where the isostatic term is small, while the recorded negative departures in recorded sea level are fairly large. The recorded and isostatic deviations agree in sign along the northeastern coast of Australia and around Japan, but the recorded departures are considerably larger than the isostatic ones. Finally, in the border area between the United States and Canada and on the tropical island Kwajalein the recorded values are positive, while the sterics are negative.

Although a number of the discrepancies may be the consequence of inadequacies in the data, the differences between the recorded and the isostatic amplitudes appear to be well established and real, probably reflecting non-isostatic processes operative in these regions.

(2) Fig. 23. May, *June*, July. The principal feature of the general isostatic pattern during this season is an extensive region of negative departures covering the central parts of the Pacific Ocean from the Aleutians to the latitude of about $20°S$. This feature is in

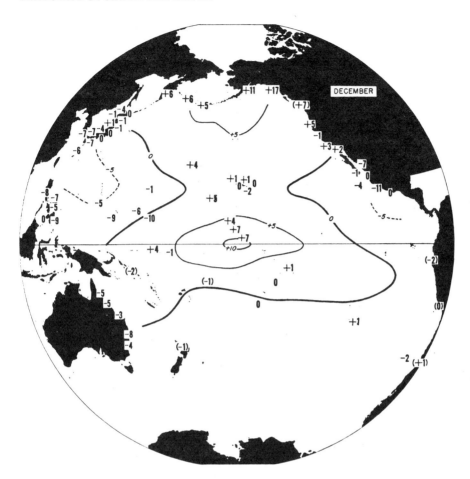

Fig. 25. Same as Fig. 22 for November, December and January. (Lisitzin and Pattullo, 1961; copyright by American Geophysical Union.)

keeping with the sea-level deviations as recorded during the International Geophysical Year. In conformity with the preceding season, the recorded sea-level departures along the coast of Mexico, although agreeing in sign, are numerically larger than the isostatic ones, and sea-level records in the Gulf of Alaska show larger negative deviations than the isostatic pattern suggests. In addition, some of the island stations do not correspond to the isostatic pattern and deviate from the recorded data of other islands in the same area. For instance, mention may be made of Johnston with a departure of −9 cm and of Wake with −4 cm. It is probable that these discrepancies are caused principally by the comparison of non-synoptic data. Sea-water temperature in the vicinity of Johnston was lower in June 1958 than during the years on which the average isostatic pattern is based. Should the temperature values for 1958 have been utilized for the determination of the steric level in this area, the isostatic departure would reach a higher negative value, resulting in a better agreement with the recorded sea-level departures. In addition, the distribution of the atmospheric pressure during the International Geophysical Year was not identical with the average for this season, which, at least to some degree, may account for the deviation noted at Wake.

The discrepancies along the coast of Mexico and Alaska, however, are probably the consequence of non-isostatic deviations of sea level.

(3) Fig. 24. August, *September,* October. On the whole, the results obtained for the two preceding figures may be repeated for this chart. The greatest discrepancies occur in the vicinity of the North American coast.

An extensive area with negative departures along the coast of British Columbia is assuredly not in isostatic equilibrium. Both the pressure term and the density term are relatively small in this region during the season, but they result, in fact, in a weak positive departure which is seen as close to the coast as the steric values can be determined. On the contrary, the observed departures along the Mexican coast are in sign in keeping with the isostatic values, but numerically too large.

In the Line Islands area the discrepancies are pronounced, the observed departures being +7 cm for Fanning and +8 cm for Palmyra. Steric data for the International Geophysical Year give −5 cm for Fanning and +3 cm for Palmyra. The smoothed isostatic contours for the region show as a whole that negative departures should be expected. Again the deviation for Johnston Island is numerically too large, +9 cm, but it is in rather good agreement with the earlier recorded average of +7 cm covering a 5-year period. One more marked discrepancy occurs at Guam, which in the chart in Fig. 24 is given as +7 cm and is taken from the results of the International Geophysical Year, while both the recorded and the steric departures during earlier periods were less than +2 cm.

(4) Fig. 25. November, *December,* January. The conformity between the isostatic pattern and the recorded values is on the whole satisfactory. The islands Kwajalein and Moen show the largest deviations with −10 cm and −9 cm respectively. The real variability is probably large in this region.

The chart shows an interesting feature for the Hawaiian Islands: an increase from the southeast to the northwest of 3 cm on the main islands alone. This increase indicates that the contribution of different small-scale or local effects may be significant even around island groups in the lower mid-latitudes.

In the southern parts of the Pacific Ocean and along the coast of South America some of the recorded values disagree in sign with the isostatic pattern, but these values are numerically rather small. It is not excluded that the zonal zero line varies fairly markedly in position, with the consequence that extensive areas may show in turn positive and negative values of departures.

In summary it may be mentioned that the recorded seasonal departures in sea level over the open parts of the ocean generally correspond in sign and in approximate magnitude with the isostatic departures determined on the basis of the variations in atmospheric pressure and water density, or specific volume. The conclusion may therefore be drawn that the large-scale changes in sea level are isostatic, which, of course, indicates that the total variations in mass of air and water per unit area are small. North of the latitude of 40°N the atmospheric fluctuations are large and consequently there must occur considerable seasonal changes in the mass of the water, whereas in the lower latitudes only the water volume, but not the water mass, varies more markedly.

Considering next the continental coasts, it may be established that conditions there are as a rule not isostatic. The most pronounced discrepancies may be found for the China Sea, the northern parts of Australia, British Columbia and Mexico south of Baja California. It is an interesting fact that in many cases, but not in all, the recorded departures are of the same sign as the isostatic term. This implies that the seasonal variation in sea level is augmented by a local coastal factor, probably the wind-stress effect.

The oceanic areas around the tropical islands frequently show fairly large discrepancies. Some of these discrepancies are due to the introduction of the average isostatic pattern in order to compare it with sea-level heights recorded during the International Geophysical Year. In these complex areas small-scale effects may not be left out of consideration. It is, in addition, not excluded that a number of the noted changes are not isostatic, and that an explanation must be sought elsewhere.

Recently Gill and Niiler (1973) performed an interesting investigation on the seasonal variability in the oceans. In their study the authors computed the seasonal variation in sea level in the North Atlantic and the North Pacific Oceans for areas situated between 15°N and 50°N, (55°N in the North Atlantic), leaving, however, the coastal regions out of consideration and paying no attention to the islands and their perturbating effects. The computed values refer to areas covering in each case 5 latitudinal and 5 longitudinal degrees. There is one more difference in the procedure if compared with that followed by Lisitzin and Pattullo. The results given by Gill and Niiler refer to the months December to February for winter, March to May for spring, etc., while Lisitzin and Pattullo based their computations on the seasonal division November to January for winter, February to April for spring, etc. Although the scheme adopted by Gill and Niiler is logically justifiable, the division chosen by Lisitzin and Pattullo corresponds more closely to the particular phases of the seasonal sea-level cycle and is therefore preferable in many respects. It must always be remembered when comparing the results that they must be affected by the deviation in period.

The most significant contribution of Gill and Niiler to the solution of the problem of the causes of the seasonal variation in sea level is the introduction of sea-surface wind stress and the response of the sea level to the seasonally variable part of this factor. The contribution of the wind stress in determining the seasonal sea-level variations in the middle parts of the oceans has so far been treated rather lightly. Veronis and Stommel (1956) computed the effect of the global-scale weather patterns upon a two-layer ocean and drew the conclusion that besides the isostatic response there should be also a barotropic response. According to Gill and Niiler, the response of the sea surface to the variable winds is mainly barotropic in the high latitudes, but becomes increasingly baroclinic towards the equator.

In their paper, Gill and Niiler determined the seasonal variations in sea level which are the consequence of the response of the water surface to the atmospheric pressure and the seasonal expansion or contraction of the upper 200-m layer due to changes in tempera- ture and salinity. Moreover, the authors computed the barotropic term representing a change of pressure at all levels. This term is estimated on the basis of wind-stress data and bottom topography. Finally, a baroclinic term which represents the effect of expansion or contraction below the depth of 200 m was computed, estimated from wind stress and density data.

In spite of largely deviating approaches to the problem, the results achieved by Gill and Niiler show a good agreement with the results given by Lisitzin and Pattullo. Thus the signs of sea-level departures determined by Gill and Niiler are, as a rule, opposite for winter and for summer. A similar feature is also characteristic of spring and autumn. Numerically the values representing the same area differ, however, quite considerably. There is also a pronounced similarity in the general pattern of the results achieved by Gill and Niiler in comparison with those given in the charts of Fig. 22–25. The conclusion drawn by Lisitzin and Pattullo, that the sea-level changes in the open parts of the oceans are mainly isostatic, implies that the additional term is small. The fact that this term has actually been calculated by Gill and Niiler and was shown to be small puts this hypothesis on a much stronger foundation. Nevertheless, the consideration of the effect of wind stress on the sea level is without doubt a considerable improvement on the older results.

According to E. Palmén (personal communication, 1972) the introduction of the sea- sonal variation of wind stress as an important factor is in no real contradiction to the isostatic theory used by Pattullo et al. The difference is, however, connected with the physical-dynamic interpretation of the ultimate causes of the sea-level fluctuations asso- ciated with the density changes of the upper layer of the oceans. These changes are, of course, largely determined by the seasonal heat changes of this layer, which are due to variations in heat radiation and exchange of sensible and latent heat between the sea surface and the atmosphere. However, the seasonal variation of the wind stress also has to be considered as causing variations in the density distribution in the upper oceanic layer, especially in the depth of the oceanic thermocline. This would imply that in computing the steric sea-level heights it is not quite correct to consider only the temperature, or density, variations in the upper 100 m layer, but that the seasonal fluctuations of the thermocline, largely caused by the variations of the wind stress, should also be taken into account. By using a greater depth for the upper 'disturbed' layer, the principle of the term steric sea level, introduced by Pattullo et al., could better include the effect of the response of the oceans to the variations in the atmospheric circulation.

In connection with the studies of the seasonal cycle in sea level on a world-wide scale, it must also be mentioned that E. Palmén (personal communication, 1972) has pointed out that the seasonal variation in sea level may be influenced by seasonal changes in the strength, and consequently also in the water transport, of the great current systems in the oceans, associated with the variation of the wind stress. The deflecting force of the Earth's rotation (Coriolis force) could contribute considerably to the variations in sea level in- and outside the great gyres. This is assuredly a significant problem which has so far not been investigated, but to which special attention should be paid in future research on the seasonal sea-level variations on a global scale.

In addition to the studies referring to the annual fluctuations in sea level with a world-wide coverage of the oceans, an appreciable number of investigations have more restricted regions as research areas. A marked variety of the principal causes of seasonal variations may easily be noted in these studies. Since it is hardly possible to pay attention to all these papers, only some of the more interesting may be referred to here.

Galerkin (1960), in a paper on the physical basis of the forecast of the variations in sea level in the Sea of Japan, summarized his results as follows:

(1) The principal cause of the seasonal changes in mean sea level in the concerned sea basin is the variation in the water density in the active layers of the sea.

(2) The effect of air pressure is also considerable.

(3) The immediate effect upon the sea surface of monsoon winds circulating over the sea basin is not marked.

The principal results achieved by Galerkin, which were in conformity with those given in older investigations, were confirmed by Lisitzin (1967b). A characteristic feature which may be of interest is the seasonal variation of the slope of the water surface in the Sea of Japan. This variation is reflected in a pronounced anticlockwise rotation, since the water surface in the Sea of Japan slopes in the winter upwards toward the northeast, in spring time toward the northwest, during summer months toward the southwest and, finally, in autumn toward the southeast.

The water interchange between the Sea of Japan and adjoining basins has not been examined on a larger scale. For the Arctic Ocean the in- and outflow of the water masses seems, in addition to atmospheric pressure, to be a significant factor in the seasonal fluctuations in mean sea level. It may, however, suffice to refer in this connection to two papers on this subject by Lisitzin (1964c, 1969).

The marked variety and complexity of the system of factors affecting the mean sea level during the particular seasons has been stressed by Galerkin et al. (1962) in a study of Indonesian waters. On the basis of a survey covering 45 stations, the authors were able to draw the conclusion that the seasonal cycle in sea level in this area is in keeping with the seasonal changes of the prevailing surface currents. It must, however, be kept in mind in this respect that the changes of the currents may be related to meteorological conditions and that these changes may influence the water volume as well.

An additional aspect of the principal elements influencing the mean sea level was given by Waldichuk (1964) in a paper on the seasonal sea-level variations on the Pacific coast of Canada. For the stations situated in the Strait of Georgia, which separates Vancouver Island from the mainland, a marked effect of river discharge has been observed. The summer maximum in sea level must be related to the pronounced spring and early summer run-off of the Fraser River. Nevertheless, according to Waldichuk is is not only for the stations situated in the Strait of Georgia, but also for other stations on the Pacific coast of Canada, that the contribution of atmospheric pressure and water density to the mean sea level may by no means be neglected.

The effect of atmospheric pressure upon the seasonal cycle in sea level in the Pacific Ocean was studied by Galerkin (1963) in more detail. Atmospheric pressure and its variations form, as has been pointed out above, only one of the factors influencing the seasonal cycle in sea level in the oceans. The results obtained by Galerkin are, nevertheless, of great interest. The most important of these results are reproduced in the following, where the mean sea level of the Pacific Ocean as a whole and for its northern and southern parts, is given for the months January and July relative to the mean sea level of the total world oceans:

	January	July
The whole Pacific Ocean	1.33 cm	1.56 cm
The northern part of the ocean	0.32 cm	0.21 cm
The southern part of the ocean	2.15 cm	2.67 cm

According to the above data, the mean sea level in the northern part of the Pacific Ocean decreases as a consequence of the effect of atmospheric pressure from January to July by 0.11 cm, while the southern part of the ocean is characterized by an increase in sea level of 0.52 cm. This value, reflecting the interchange of water masses between the Pacific Ocean and the adjacent oceans, it thus comparatively slight. It seems therefore reasonable that no attention should be paid to it in connection with computation, for instance, of the water balance of the Pacific Ocean.

THE SEASONAL VARIATION OF THE SLOPE OF THE WATER SURFACE

The problem of the seasonal variation of the slope of the water surface is closely related to the question of the seasonal cycle of sea-level heights, since both phenomena are principally due to the contribution of meteorological and hydrographic factors. The method and the results described in the following give, in addition, the magnitude of the average slope of the water surface in the investigated basins and are connected also in this respect with the problem of determining the mean sea level in oceans and seas. Reference to the results presented below will therefore also be made in connection with this problem (pp. 143–163).

The slope of the water surface and its seasonal variations may most easily be investigated in semi-enclosed sea basins with numerous sea-level stations operating along the coasts and with a sufficient quantity of meteorological and hydrographic data. The Baltic Sea and its large bays, the Gulf of Bothnia and the Gulf of Finland correpond fairly well to these requirements. The results obtained for these sea basins will therefore be described below.

Witting (1918) was the first to pay attention to the question of the slope of the water surface in the Baltic Sea. However, the results of his comprehensive studies were published more than half a century ago, before the present extensive network of sea-level recording stations was developed in the area. Some 40 years later the situation had changed considerably. There were, for instance, practically uninterrupted series of records available covering more than 30 years for 9 of the 10 Finnish sea-level gauges in the Gulf of Bothnia and adjacent regions. Moreover, hydrographic data on water temperature and salinity made in the open sea during research cruises had increased and reliable observations on atmospheric pressure and wind were available. A new attempt to determine the slope of the water surface in the Gulf of Bothnia and its seasonal variation seemed therefore to be appropriate (Lisitzin, 1957b).

According to the results obtained by Witting, the mean sea level in the innermost parts of the Gulf of Bothnia stands, on average, as a consequence of decreasing density of the water, approximately 30 cm higher than the sea level at the position $57°N$ $3°E$ in the North Sea, where salinity is supposed to be 35 ‰ . The largest part of this increase in sea-level height occurs in the transition area between the North Sea and the Baltic. In the latter basin the increase is only approximately 10 cm. Of this amount less than half, or some 4 cm, may be ascribed to the Gulf of Bothnia.

For the determination of the slope of the water surface brought about by differences in density, the choice of the reference level is very important, since it should represent the level of no motion, or at least the level where pressure differences should be at a minimum. Dietrich (1954) basing his results on a research performed by Jacobsen (1943) assumed that the hydrostatic pressure close to the bottom is identical to that prevailing in the water layers of adjoining deeper areas at the corresponding depth. In the Baltic Sea, where the average depth is approximately 60 m and the topography of the bottom fairly complicated, the determination of the level of no motion is still more complex. Witting tried to solve this problem by basing the computations on different reference levels for different seasons.

For the determination of the results presented below, the depth of 20 m was chosen as reference level. A number of objections may be raised against such a choice, but it can be justified from other points of view. Firstly, according to Granqvist (1938) the seasonal variation in salinity is not pronounced in the Gulf of Bothnia, and the increase in salinity with the depth is relatively weak in this basin. The consequence is that the seasonal changes in water density are mainly dependent upon the corresponding variation in temperature. Secondly, the data giving the increase in sea level due to decreasing water density are compared with the recorded data of the gauges which are situated along the coast, where the depth is at the utmost some 20 m.

The relative differences in the height of the sea level caused by differences in water density were computed from the data on temperature and salinity collected in the open sea during hydrographic expeditions covering 25 years and referring to the months May, July and November. These data were averaged by Granqvist (1938). The results are set out in Table XXIV. In this table the sea level at the hydrographic station F68, situated in the southernmost part of the area under investigation, was chosen as the zero level. The position of the hydrographic stations from which data have been used is shown in Fig. 26.

Table XXIV shows that the selection of the reference level of 20 m for the Gulf of Bothnia gives satisfactory results. The deviations in the slope of the water surface in the

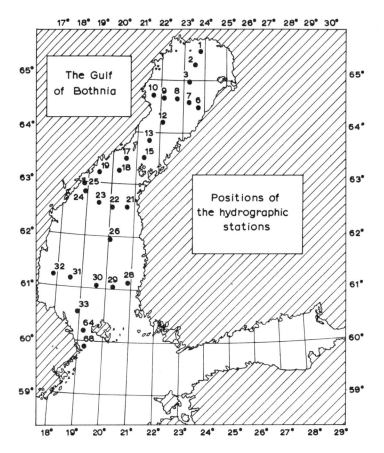

Fig. 26. Positions of hydrographic stations, Gulf of Bothnia.

open sea are not pronounced during the three months. There are only three cases in the table for which a particular value has deviated by more than 0.5 cm from the average for the station. These cases refer to the stations F18 and F25A located in the Kvark or its immediate vicinity, i.e., in a region characterized by a marked density decrease. It may thus be assumed that the slope remains practically unchanged during the whole year. The average height difference between the northernmost parts of the Gulf of Bothnia and the hydrographic station F68 amounts to 4.7 cm, a result corresponding rather closely to the value computed by Witting.

The average monthly differences in atmospheric pressure (in mbar) between the southern and northern parts of the Gulf of Bothnia are given in Table XXV. The data in this table refer to the relevant differences between Stockholm and Haparanda for two periods and between Mariehamn and Marjaniemi for the shorter of these periods. The longer period, 1926–1955, coincides with the period of sea-level records on which the following

TABLE XXIV

RELATIVE DIFFERENCES IN SEA LEVEL (CM) CAUSED BY DECREASING DENSITY IN THE GULF OF BOTHNIA

Station	May	July	November	Mean	Station	May	July	November	Mean
F1	4.8	5.1	4.2	4.7	F21	0.2	0.1	−0.1	0.1
F2	3.6	3.9	3.4	3.6	F22	0.2	0.1	0.1	0.1
F3	3.6	3.5	3.6	3.6	F23	0.4	0.4	0.3	0.4
F6	3.4	3.1	3.2	3.2	F24	0.6	1.0	0.3	0.6
F7	3.4	3.2	3.3	3.3	F25A	0.4	1.4	0.3	0.7
F8	3.5	3.3	3.3	3.4	F26	0.0	−0.1	0.0	0.0
F9	3.4	2.9	3.2	3.2	F28	0.0	−0.2	−0.2	−0.1
F10	3.7	3.8	3.3	3.6	F29	0.0	0.1	−0.1	0.0
F12	3.3	3.1	3.2	3.2	F30	0.0	0.0	−0.2	−0.1
F13	3.2	3.4	3.2	3.3	F31	0.2	0.0	0.2	0.1
F15	2.9	3.1	2.7	2.9	F32	0.3	0.3	0.4	0.3
F17	2.7	2.9	2.1	2.6	F33	0.2	0.5	0.4	0.4
F18	1.0	1.5	0.0	0.8	F64	0.0	0.3	0.0	0.1
F19	0.7	1.2	0.2	0.7	F68	0.0	0.0	0.0	0.0

results have been based, while the shorter period is given in order to confirm the reliability of the atmospheric pressure data. A comparison of the particular values in Table XXV indicates that for the determination of the seasonal variation of the slope of the water surface a possible inaccuracy of approximately 0.3 mbar for each month may have to be reckoned with.

Comparing the general features of the seasonal cycle of differences in atmospheric pressure for the different stations and periods, a rather pronounced secondary maximum may be noted in June. This maximum is interesting in this respect since its effect is distinctly reflected in the sea-level records.

The most significant factor contributing to the seasonal fluctuations of the slope of the water surface in the Gulf of Bothnia is, however, the piling-up effect of the wind. In computing this effect the formula given by Palmén (1936) has been used. This formula

TABLE XXV

AVERAGE MONTHLY DIFFERENCES OF ATMOSPHERIC PRESSURE (MBAR) BETWEEN THE SOUTHERN AND NORTHERN PARTS OF THE GULF OF BOTHNIA

	J	F	M	A	M	J	J	A	S	O	N	D	Year
Difference Stockholm−Haparanda													
1926−1955	2.0	2.2	2.8	1.8	0.1	1.2	0.3	0.7	2.4	2.8	2.7	2.2	1.8
1921−1940	1.6	2.4	2.4	0.0	−0.1	1.1	0.0	0.1	2.4	2.4	2.4	2.3	1.4
Difference Mariehamn−Marjaniemi													
1921−1940	1.7	1.6	1.7	0.1	−0.4	0.9	0.0	0.3	1.8	1.8	2.0	2.0	1.1

had to be modified to some degree, since it was deduced originally for an actual case and not for the evaluation of monthly averages. The formula used was:

$$\Delta H = \frac{3.2\,\alpha\,L\,v^2}{d}$$

where ΔH is the difference in sea-level heights in cm for a horizontal distance L, expressed in 100 km, d the average depth in m of the sea area to be examined, and v the average monthly velocity of the wind in m/sec. α is a factor indicating the relative frequency of different wind directions. As a rule, the slope existing in the Gulf of Bothnia as the consequence of the density gradient of the sea water and the differences in atmospheric pressure is reinforced by southerly winds and weakened by northerly winds. If n_S denotes the relative number of winds covering the former group and n_N the corresponding number for the latter group, the relation:

$$\alpha = \frac{n_S - n_N}{n_S + n_N}$$

may be used as the factor indicating the changes of the wind effect during particular months.

As soon as the contribution of the meteorological and hydrographic factors was determined, the seasonal variation of the slope of the water surface between the separate sea level stations in the Gulf of Bothnia and Degerby situated in the southern part of the Aaland Islands could be determined. Fig. 27 gives the results for the northernmost station Kemi and Degerby. This figure shows very distinctly that the contribution due to the density effect varies only slightly in the course of the year, and that the effect of

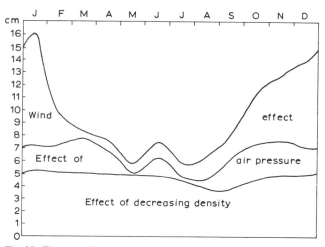

Fig. 27. The contribution of density, atmospheric pressure and wind to the relative heights of sea level between Kemi and Degerby (Lisitzin, 1957b).

atmospheric pressure is the weakest of the three effects, but that the deviations for the separate months are in this case relatively marked. The latter feature is still more accentuated in connection with the wind effect, being, for instance, in January 18 times larger than in March. During the months December and January the wind contributes with approximately 50% of the total height difference in sea level between Kemi and Degerby.

The sea-level data which were used in order to compare with the results determined on the hydro-meteorological basis refer, as already mentioned, to the 30-year period 1926–1955. The seasonal variations in sea level have for every station been computed with respect to the mean sea level for the whole period. Since the heights are related to the mean sea level of the particular stations, it is not possible to determine the slope of the water surface during different months directly from the data. However, the data allow some general conclusions to be drawn. For instance, the sea level was at Kemi 4.8 cm higher in January than in July, while at Degerby the opposite feature could be noted, the sea level being 2.9 cm lower in January than in July. This indicated that the height difference from the north to the south in the Gulf of Bothnia decreased during half a year by 7.7 cm. In order to obtain the relative heights in sea level at the particular stations for the different months, the heights were determined in relation to Degerby. The results for the three stations Kemi, Pietarsaari and Kaskinen are reproduced in Table XXVI. Of special interest in this connection is the secondary maximum in June which occurs simultaneously with the secondary maximum for the difference in atmospheric pressure in Table XXV.

At this stage in the study it was possible to compare the hydrographic and meteorological contribution to the sea level with the recorded data. The three Tables XXVII–XXIX give the results. The first line in these tables reproduces the sea-level differences caused by water density. It may be pointed out here that these data are the least reliable in the tables, since they are based on observations for three months only. The second line shows the relative effect of the atmospheric pressure, interpolated for the particular stations on the basis of Table XXV. The third line gives the differences in sea-level heights due to the piling-up effect of the wind. The next line represents the total heights as the result of the three hydro-meteorological factors. The fifth line gives the corresponding data from Table XXVI, to which the average yearly effects of density, atmospheric pressure and wind have been added. Finally, the last line in each table represents the deviations between the results achieved by the two independent methods used for the determination of the seasonal cycle in sea-level differences in the Gulf of Bothnia.

TABLE XXVI

RELATIVE SEA-LEVEL HEIGHTS (CM) WITH THE SEA LEVEL AT DEGERBY AS REFERENCE LEVEL

	J	F	M	A	M	J	J	A	S	O	N	D
Kemi	3.6	0.1	−1.6	−2.0	−4.5	−2.5	−4.1	−3.0	−0.6	1.7	5.4	6.6
Pietarsaari	2.6	0.6	−0.6	−0.6	−3.0	−1.3	−3.0	−2.1	−0.6	0.9	2.9	3.7
Kaskinen	2.0	0.2	−0.4	0.0	−1.7	−0.8	−1.8	−1.4	−0.6	0.1	1.5	2.3

The seasonal variations in height differences for all Finnish sea-level stations situated in the Gulf of Bothnia are reproduced graphically in Fig. 28. This figure shows, in accordance with Tables XXVII—XXIX that the largest positive deviations, especially pronounced for the northernmost stations, occur in January, and the greatest negative deviations generally in November and December. This result is by no means surprising, since the wind effect is relatively most marked during these months, and small changes in the average wind velocity may affect the deviations considerably. Moreover, it must be kept in mind that the final results are in some cases based on approximate data. The positive deviations in January may be ascribed, at least to some degree, to the occurrence of the continuous ice cover, which reduces the piling-up effect of the wind (pp. 80—86). Never-

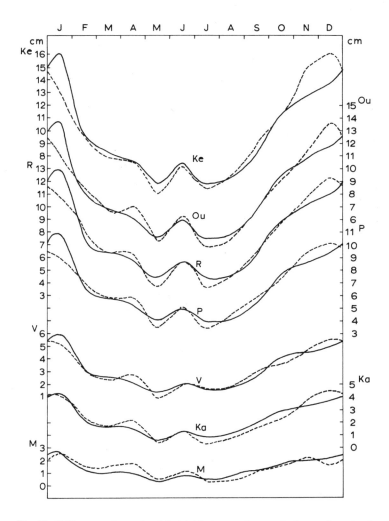

Fig. 28. Differences in sea-level height between the particular stations in the Gulf of Bothnia and Degerby. The solid lines are determined with the help of the hydrographic and meteorological data; the dashed curves on the basis of sea-level records (Lisitzin, 1957b).

TABLE XXVII

SEA-LEVEL DIFFERENCES (CM) BETWEEN KEMI AND DEGERBY

	J	F	M	A	M	J	J	A	S	O	N	D	Year
Density	5.2	5.1	5.0	4.9	4.9	4.8	4.5	3.8	3.8	4.5	4.8	4.9	4.7
Atm. pressure	2.0	2.2	2.8	1.8	0.1	1.2	0.3	0.7	2.4	2.8	2.7	2.2	1.8
Wind	8.8	2.4	0.5	0.9	0.8	1.4	1.0	1.8	1.9	4.0	5.2	6.7	3.0
Total	16.0	9.7	8.3	7.6	5.8	7.4	5.8	6.3	8.1	11.3	12.7	13.8	9.5
Sea level	13.1	9.6	7.9	7.5	5.0	7.2	5.4	6.5	8.9	11.2	14.9	16.1	9.5
Difference	2.9	0.1	0.4	0.1	0.8	0.2	0.4	-0.2	-0.8	0.1	-2.2	-2.3	—

TABLE XXVIII

SEA-LEVEL DIFFERENCES (CM) BETWEEN PIETARSAARI AND DEGERBY

	J	F	M	A	M	J	J	A	S	O	N	D	Year
Density	3.7	3.6	3.5	3.4	3.4	3.3	3.0	3.3	3.3	3.0	3.3	3.4	3.2
Atm. pressure	1.3	1.5	1.9	1.2	0.1	0.8	0.2	0.8	0.5	1.9	1.8	1.5	1.2
Wind	5.9	1.6	0.3	0.6	0.5	0.9	0.7	0.9	0.9	2.7	3.5	4.5	2.0
Total	10.9	6.7	5.7	5.2	4.0	5.0	3.9	5.0	5.0	7.6	8.6	9.4	6.4
Sea level	9.0	7.0	5.8	5.8	3.4	5.1	3.4	5.1	5.8	7.3	9.3	10.1	6.4
Difference	1.9	-0.3	-0.1	-0.6	0.6	-0.1	0.5	-0.1	-0.5	0.3	-0.7	-0.7	—

TABLE XXIX

SEA-LEVEL DIFFERENCES (CM) BETWEEN KASKINEN AND DEGERBY

	J	F	M	A	M	J	J	A	S	O	N	D	Year
Density	0.3	0.3	0.3	0.3	0.3	0.3	0.3	0.2	0.2	0.3	0.3	.0.3	0.3
Atm. pressure	0.8	0.9	1.1	0.7	0.0	0.5	0.1	0.3	1.0	1.1	1.1	0.9	0.7
Wind	3.2	0.9	0.2	0.4	0.3	0.5	0.4	0.6	0.7	1.5	1.9	2.5	1.1
Total	4.3	2.1	1.6	1.4	0.6	1.3	0.8	1.1	1.9	2.9	3.3	3.7	2.1
Sea level	4.1	2.3	1.7	2.1	0.4	1.3	0.3	0.7	1.5	2.2	3.6	4.4	2.1
Difference	0.2	-0.2	-0.1	-0.7	0.2	0.0	0.5	0.4	0.4	0.7	-0.3	-0.7	—

TABLE XXX

SEA-LEVEL DIFFERENCES (CM) BETWEEN DEGERBY AND GEDSER

	J	F	M	A	M	J	J	A	S	O	N	D	Year
Hydro-meteorological total	22.4	13.3	11.6	9.8	6.3	13.1	11.6	13.5	14.3	16.1	16.4	19.5	14.0
Sea level	20.0	13.1	10.9	10.6	3.8	11.7	13.7	13.5	15.4	18.1	17.9	21.6	14.0
Difference	2.4	0.2	0.7	−0.8	2.5	1.4	−2.1	0.0	−1.1	−2.0	−1.5	−2.1	–

theless, leaving all these factors out of consideration, it may be noted that the average deviations are not marked, particularly in comparison with the range of the monthly sea-level changes. The following values may be of interest in this connection:

	Range of the variations (cm)	Mean deviation (cm)
Kemi	11.1	1.3
Pietarsaari	6.7	0.7
Kaskinen	4.1	0.4

Following the same procedure as described above, the slope of the water surface was determined for the Gulf of Finland (Lisitzin, 1958a) and for the Baltic proper (Lisitzin, 1962). For the Gulf of Finland the following results were achieved: sea-level height difference between Hamina, situated in the eastern part of the Gulf of Finland and Degerby has, computed on the basis of the meteorological and hydrographic data, resulted in the yearly average value of 5.0 cm. Of the total contribution 3.7 cm must be ascribed to the density effect and 1.3 cm to the piling-up effect of the wind stress. On average there is no contribution of atmospheric pressure. As a whole, the results for the Gulf of Finland were somewhat less satisfactory than for the Gulf of Bothnia, the mean deviations reaching relatively higher values.

The computations for the Baltic Sea proper, covering the region between Degerby and Gedser in Denmark, have shown that the average height differences in sea level due to the hydro-meteorological effects may be estimated at 14.0 cm. The largest part of this contribution, 7.9 cm, is due to the density effect, 1.8 cm may be attributed to the effect of atmospheric pressure and 4.3 cm is the consequence of the wind-produced piling-up of the water. In this respect it may be mentioned that according to the computations by Witting (1918), the total anomaly in height amounts to 15 cm. The results obtained by Lisitzin are reproduced in Table XXX. The correspondence between the hydro-meteorological contribution and sea-level records is for the Baltic proper more satisfactory than for the Gulf of Finland, but not as pronounced as the case for the Gulf of Bothnia.

THE SEASONAL WATER BALANCE OF THE OCEANS

The determination of the seasonal balance of water in the oceans is a problem which may be of considerable interest not only for oceanographers, but also for hydrologists and meteorologists. It is one of those numerous problems where to find a solution sea-level specialists have to collaborate with and make use of the results of scientists representing many different branches of geophysics. In this connection it may be significant to point out that the total quantity of water participating in the seasonal interchange between oceans, continents and atmosphere may be estimated at $50 \cdot 10^{19}$ g (or cm^3). This water quantity corresponds to 0.3‰ of the water volume in the oceans and seas, and to 2% of the landbound part of the hydrosphere. Finally, the amount of water participating in the seasonal interchange is approximately 40 times as large as the average quantity of water vapour in the atmosphere.

Munk (1958) made the first attempt to investigate the extent to which the total water masses in the oceans are subject to seasonal variations. Munk's best estimate was that between October and March oceans in the northern hemisphere lose $2 \cdot 10^{19}$ g of water, while oceans in the southern hemisphere gain $1 \cdot 10^{19}$ g, resulting thus in a net loss of $1 \cdot 10^{19}$ g. However, Munk himself has pointed out that the inaccuracy of the estimate was such that the loss might be twice as large, or there might be no loss at all. In addition, Munk expressed the expectation that sea-level observations made during the International Geophysical Year 1957–1958 would contribute to narrow the uncertainty a great deal.

This expectation has been justified. It was on the basis of sea-level records for the International Geophysical Year and the subsequent years of International Collaboration that Lisitzin (1970) obtained the results concerning the seasonal water balance in the oceans. The five-year period 1957–1961 was used for this purpose in spite of the fact that the records for the first and especially for the last year of the series were less complete than for the remaining years. It was therefore necessary to inter- and extrapolate the data over extensive areas of the oceans. This applies in particular to the Arctic Ocean, to the Antarctic sea regions and, in numerous cases, to the deep parts of the open oceans. The method of computation was rather elementary. The increase or decrease in sea level from March to September was determined separately for the two hemispheres and for the entire area covered by sea water. The results are collected in Table XXXI, which gives not only the changes in sea level, but also the variations in water quantity. Studying this table more closely, it may be immediately noted that the variability is highly pronounced for particular years. This is partly the consequence of the variability of the relevant phenomenon itself, also partly depending, however, upon the inadequate inter- and extrapolation of sea-level data over extensive oceanic areas. A characteristic feature, noted in all years, is the fact that the increase in sea level in the northern hemisphere is more accentuated than the corresponding decrease in the southern hemisphere. On average, the increase in sea level is in the northern hemisphere 7.6 cm and the increase in water quantity $1.19 \cdot 10^{19}$ g, while the corresponding decreases in the southern hemisphere are 3.5 cm and $0.71 \cdot 10^{19}$ g, resulting thus in a net gain of 1.3 cm and $0.48 \cdot 10^{19}$ g.

Before proceeding to explore the question of the origin of the actual gain of water in the oceans during the period from March to September, two factors must be taken into consideration. The gain in sea water in the northern hemisphere and the corresponding loss in the southern hemisphere are to a certain degree due to the changes in volume as a consequence of heating and cooling respectively of the water masses in accordance with the shift of the seasons. The thermal contribution to the water volume was therefore computed on the basis of available oceanographic data. The results were as follows: in the northern hemisphere the thermally caused increase in sea level from March to September could be estimated at 4.6 cm and $0.71 \cdot 10^{19}$ cm^3 (or g), while in the southern hemisphere the corresponding decrease amounts approximately to 4.0 cm and $0.82 \cdot 10^{19}$ cm^3 (or g).

TABLE XXXI

SEASONAL CHANGES (MARCH TO SEPTEMBER) IN SEA LEVEL AND WATER QUANTITY

	1957	1958	1959	1960	1961	1957–1961	Density effect	Residue	Air pressure effect	Final residue
Changes in sea level in cm										
Northern hemisphere	6.2	10.5	8.3	5.7	7.4	7.6 ± 0.9	4.6	3.0	−1.5	1.5
Southern hemisphere	−1.7	−4.0	−4.6	−3.3	−3.7	−3.5 ± 0.5	−4.0	0.5	1.1	1.6
Total	1.7	2.4	0.9	0.6	1.1	1.3 ± 0.3	−0.3	1.6	–	1.6
Changes in water quantity in 10^{19} g										
Northern hemisphere	0.96	1.67	1.29	0.88	1.15	1.19 ± 0.14	0.71	0.48	−0.23	0.25
Southern hemisphere	−0.35	−0.82	−0.95	−0.68	−0.76	−0.71 ± 0.10	−0.82	0.11	0.23	0.34
Total	0.61	0.85	0.34	0.20	0.39	0.48 ± 0.11	−0.11	0.59	–	0.59

Deducting these values from the results giving the average sea-level changes during the five-year period, we obtain the values which are presented in the column denominated 'Residue' in Table XXXI. In the northern hemisphere there is an increase of 3.0 cm or $0.48 \cdot 10^{19}$ g and in the southern hemisphere an increase of 0.5 cm or $0.11 \cdot 10^{19}$ g. For the entire area covered by sea water the increase amounts to 1.6 cm or $0.59 \cdot 10^{19}$ g.

The increase in the water quantity is thus several times larger in the oceans of the northern hemisphere than in the oceanic regions of the southern hemisphere. This discrepancy is the consequence of the shift of the distribution of atmospheric pressure over the oceans from March to September. Taking into account that atmospheric pressure decreases during this period over the oceans in the northern hemisphere and increases over the sea-water-covered regions of the southern hemisphere, and that considerable air masses are transported towards the continents in winter, it can be estimated that the water quantity passing across the equator from the south to the north during the half-year period from spring to autumn amounts to $0.23 \cdot 10^{19}$ g (Lisitzin, 1960). The final residue, excluding the changes occurring in the oceans themselves, would thus result in an approximate increase of 1.5 cm or $0.25 \cdot 10^{19}$ g, in the oceans of the northern hemisphere and in 1.6 cm, or $0.34 \cdot 10^{19}$ g, in those of the southern hemisphere. The results for the two hemispheres are thus in reasonable agreement. Nevertheless, because of the fairly pronounced mean deviations, it is not possible to ascribe too much significance to this favourable result.

The total increase of the water quantity in the oceans from March to September, $0.59 \cdot 10^{19}$ g, agrees fairly well with the results presented by Van Hylckama (1956), according to which the maximum detention of moisture on land in the form of snow, moisture in the soil, ground water and organic material occurs in March and April and the minimum detention in September and October. The total range of this variation was estimated by Van Hylckama at $0.50 - 0.75 \cdot 10^{19}$ g.

There is now left the interesting question of the origin of the water gained by the oceans from spring to autumn. The question may be posed as to which of the factors mentioned by Van Hylckama is the most significant for the problem as a whole. It may also be asked whether some part of the water gain in September is possibly stored in the atmosphere in March.

The interchange of water between the oceans and the continents during the half-year period from early spring to early autumn seems to be the predominant factor. It is a well-known fact that considerable snow masses cover Siberia during the winter months. Making use of a chart published by Pupkov (1964) which gives the average water values of the snow cover in this region for the 20-year period 1941–1960, the water storage in Siberia from early spring to early autumn can be estimated at $0.15 \cdot 10^{19}$ g. Attention must be also paid to other polar and subpolar regions with a seasonal snow cover. The contribution for the entire area in question was estimated by Shumski et al. (1964) in the following way. Snow precipitated on the surface of the Earth comprises $2-3 \cdot 10^{19}$ g. A considerable part of this snow, however, melts almost immediately when reaching water and warm soil. In the northern hemisphere, owing to the extensive landarea, snow covers roughly 80,000,000 km^2, whereas in the southern hemisphere the corresponding area amounts only to about 40,000,000 km^2. Snow masses have therefore a minimum volume corresponding to $0.75 \cdot 10^{19}$ g late in the northern summer and a maximum volume amounting to $1.35 \cdot 10^{19}$ g towards the end of the northern hemisphere winter. There are then $0.60 \cdot 10^{19}$ g more snow on the continents in March than in September. The assumption may be made that only 75% of the total water quantity reaches the oceans during the melting period and subsequent months before the beginning of the next winter. The final result obtained is that approximately $0.45 \cdot 10^{19}$ g of the gain in water may be considered to be a consequence of the storage of snow on the continents of the northern hemisphere in March. This value may appear comparatively high, but it has assuredly a real background, since the computations of Zubenok, quoted from Malkus (1962), have shown that the river discharge to the Arctic Sea during the summer half of the year may be estimated at $0.26 \cdot 10^{19}$ g. The discrepancy of $0.19 \cdot 10^{19}$ g may easily be accounted for by taking into consideration a large number of rivers originating in and flowing through the areas with seasonal snow cover and discharging their water masses into the Pacific or Atlantic Oceans or the Caspian Sea. In this respect it must also be taken into account that glaciers are more extensive in the southern hemisphere than in the northern hemisphere, and consequently the detention of water as ice by the continents is larger in September than in March. The melting and sublimation of continental ice may therefore to some extent counterbalance the effect of snow melting. However, the seasonal changes in the quantity of continental ice seem to be markedly less than the fluctuations in the quantity of snow. It can be estimated that the loss of water from the oceans to the glaciers from early spring to early autumn of the northern hemisphere amounts to $0.04 \cdot 10^{19}$ g. Continental snow and ice may then be considered to contribute $0.41 \cdot 10^{19}$ g of the total gain of water by the oceans, which according to the results obtained on the basis of sea-level data occurs from March to September.

The other possible source for the water gain is the atmosphere. Different estimates of the total quantity of water in the atmosphere are available. The most commonly adopted amount is probably $1.3 \cdot 10^{19}$ g and this is only two to three times as large as the relevant increase in water quantity. Some authors have assumed that the amount of water in the atmosphere may fall as low as $0.70 \cdot 10^{19}$ g. Generally, the seasonal fluctuations of water vapour in the atmosphere have been considered to be only a slight fraction of the total amount reaching approximately 10%. On the basis of the statement made by Budyko (1956) that in the southern hemisphere evaporation is 12 cm per year higher than precipitation and taking, in addition, account of the seasonal cycle in evaporation and the pronounced moisture transport across the equator which is primarily due to the Asiatic monsoon, there may be a water gain for the oceans from the atmosphere during the period from March to September. This gain may be estimated only very roughly at $0.12 \cdot 10^{19}$ g, or approximately at 20% of the total water gain.

The results obtained show that the water storage in the atmosphere as water vapour and on the continents in the form of snow and ice amounts to $0.53 \cdot 10^{19}$ g in March. Allowing for the inaccuracy involved in the particular stages of the estimations, this result is satisfactory. It must always be borne in mind that, as different authors have pointed out, the variability of the water discharge from the Siberian rivers may amount to 20%. Finally, according to Van Hylckama, there are different additional factors, such as humidity of the soil, ground water and organic material, which may contribute to the increase of moisture detention by the continents in March and April. Although the author is forced at the present time to renounce any attempt to evaluate the contribution of these factors, the assumption may be made that a water gain by the oceans of the magnitude of $0.53-0.59 \cdot 10^{19}$ g is by no means beyond the limits of possibility.

CHAPTER 5

A WORLD-WIDE MEAN SEA LEVEL AND ITS DEVIATIONS

It is general practice to choose the average physical water surface as the reference level not only for all measurements of altitude on land, but also for all depth determinations at sea. Nevertheless, every oceanographer is aware that the mean sea level in the oceans and seas varies considerably and that deviations from the average may not be neglected even for practical purposes. In order to give an example, it may be mentioned that owing to the fact that the water density deviates highly between the Atlantic Ocean and the Pacific Ocean, the latter ocean stands, on the whole, higher than the former ocean. A number of other contributing factors must also be taken into consideration. Although the effect of the wind-caused piling-up of water cannot be proved to occur in the deep open parts of the oceans, special attention must always be paid to this effect when comparing the mean sea level at two places situated at some distance from each other along the coast, especially if the orientation of the two places is different. It must also be remembered that the effect of piling-up increases in shallow water and land-locked basins.

The examples given above were selected at random, and may be sufficient as an introduction to the problem of determination of the mean sea level in the world oceans, including the adjacent and Mediterranean-type seas. Moreover, these examples show very distinctly that there exists a pronounced difference concerning the approach to the problem for the open parts of the oceans on the one hand, and for the off-shore regions on the other hand. In the former areas the computations of the dynamic topography of the water surface are the principal assets for the solution of the problem. In addition, the perturbating effect of atmospheric pressure must be taken into account and also the contribution of wind stress. The conclusion can therefore be drawn that it is oceanography, in the proper, limited sense of the term, and meteorology which contribute to the solution. On the contrary, in the near-shore regions the investigations have principally to be based upon the records of the gauges and the study of long-term observations of the effect of the meteorological elements on the variations in sea level. In coastal areas where the results of geodetic precise levelling are available, these results may be of considerable use as an auxiliary source of information. In order to achieve a world-wide coverage of the problem, basins connected with the oceans by narrow and shallow passages and characterized by water masses with a thermo-haline structure highly deviating from that in the oceans must be included in the investigations. This is probably the most difficult part of the work. In summary it may therefore be noted that the problem, as a whole, consists of three principal parts. In connection with the study distinction must be made between:

(1) the open deep regions of the oceans,

(2) the adjacent and Mediterranean-type seas and the transition areas between them and the oceans and

(3) the near-shore regions in the oceans and seas.

THE OPEN DEEP REGIONS OF THE OCEANS

In spite of the fact that the broad outlines of these investigations are distinct, the practical determination of the dynamic topography of the physical water surface in the oceans meets with several difficulties. The selection of the reference level for the dynamic computations is assuredly of fundamental significance, particularly in cases where knowledge of the dynamic relief is needed for the determination of water circulation. In connection with studies of the mean sea level, the main purpose is to determine the relative topography of the physical water surface. Different authors have chosen as reference or zero level for the computations a definite constant depth which was uniform for the whole area of the ocean or for some more restricted part of it. A number of papers based on this principle are available, but it is sufficient to refer here to a study by Dietrich (1937) in which the results of the older dynamic computations are summarized.

Other oceanographers have chosen a rather different approach to the problem, endeavouring to determine for every part of the oceans the layer of 'no motion' or, to be more exact, the layer characterized by a minimum of water motion. The determination was made on the basis of different oceanographic observations and the knowledge of the distribution of a number of chemical elements with the depth. The results obtained in this way have shown that the depth of 'no motion' varies markedly in different parts of an ocean. Nevertheless, the occurrence of a pronounced regional continuity can, as a rule, be proved. In this respect it may be pointed out that Dietrich was probably one of the first to ascertain the existence of a layer of minimum water motion. Dietrich (1937) was of the opinion that the layers of a minimum oxygen content must also correspond to the layers of minimum horizontal motion in water masses. As proof of this assumption, Dietrich mentioned poor replenishment of the oxygen gas which is consumed by biological processes, as being the consequence of weak horizontal flow.

Defant (1941) based his extensive determinations of the reference level in the Atlantic Ocean on the computed data for the anomalies in specific volume between neighbouring oceanographic stations for which data on water temperature and salinity were available. In this manner Defant was able to establish that the distance between isobaric surfaces is practically constant within some characteristic layers, and drew the conclusion that the layer of minimum water motion must be situated there. The result of Defant's investigations is the well-known chart giving the absolute dynamic topography of the physical water surface in the Atlantic Ocean (Fig. 29). This chart must even now be considered as one of the outstanding achievements in oceanographic research. It is thus not only surprising but also highly regrettable that no determinations of the absolute topography of

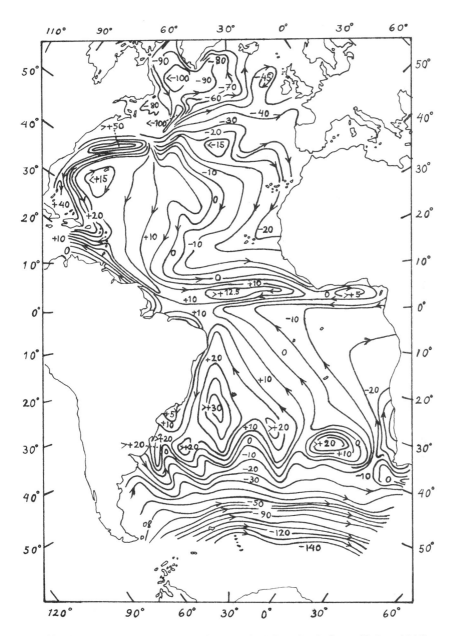

Fig. 29. The absolute dynamic topography (in cm) of the Atlantic Ocean (Defant, 1941).

the physical surfaces in the other oceans has so far been attempted on the basis of the method developed and used by Defant.

All more recent charts covering particular oceans or the world oceans as a whole refer to a uniform reference level. In this respect it may be appropriate to mention the most important of these charts. Reid (1961a) gave the topography of the physical water surface relative to the 1,000-decibar level for the Pacific Ocean. Lacombe (1951) determined the dynamic topography with respect to the 1,000-, 2,000- and 3,000-decibar surfaces for the Indian Ocean. For the waters surrounding the Antarctic continent, Deacon (1937) constructed a graphical reproduction of the dynamic topography of the physical water surface relative to the 3,000-decibar level. Although referring to a more restricted region only, the computations made by Dietrich (1935) of the physical water surface relative to the 1,000-decibar surface of the area off the southern extremity of Africa may also be listed, since this region is of special interest as a transition link between two oceans with deviating oceanographic characteristics.

Considering next the entire world ocean system, mention must be made of the studies performed by Stommel (1964, 1965) and Lisitzin (1965). The chart given by Stommel is based on earlier charts for the separate oceans and oceanic regions prepared by other authors. This chart shows the dynamic topography of the water surface relative to the 1,000-decibar level. In this connection it may be of interest to mention a few of the more characteristic features of this chart. The lowest sea-level heights, less than 0.4 dyn. m, are noted approximately at the latitude of 70°S. Proceeding towards the north the sea level increases but not, however, entirely symmetrically in particular oceans. The increase is less accentuated in the Atlantic Ocean than in the Pacific Ocean. The most pronounced sea-level heights occur, according to the results obtained by Stommel, in the western parts of the oceans; they reach the values of 1.60 dyn. m in the Atlantic Ocean, 2.00 dyn. m in the Indian Ocean and 2.20 dyn. m in the Pacific Ocean. The last-mentioned value corresponds to the greatest height in mean sea level given by Stommel and the total range of the variations is thus 1.80 dyn. m.

Lisitzin, too, has to a certain extent based her computations on the results of previous papers concerned with the problem. However, having chosen the 4,000-decibar level as reference depth, the results deviate quantitatively, if even not as markedly qualitatively, from the results presented by Stommel. The choice of the 4,000-decibar level seems to be justified by the following considerations. Since the relative height of the sea level in the particular oceans, their various basins and latitudinal belts is the central problem, the greatest possible depth should be taken into consideration. Moreover, since variations of depth in the oceans are considerable, especially if account is taken of the separate basins, deep trenches and numerous ridges, the choice is very difficult. However, the average depth of the oceans, excluding adjacent and Mediterranean-type seas, is fairly close to 4,000 m. In addition, oceanographic data reaching down to this depth are not too scarce. On the other hand, substantial parts of the individual oceans are separated from one another by regions where the depth is less than 4,000 m. This fact prevents the mixing of the heavier bottom waters in different basins. Nevertheless, where a connection was necessary between particular basins of oceans or between different oceans over more shallow areas, attention was paid to the sea-floor relief of the region and to the depth of the ridges, a procedure which has been applied more radically in connecting the mean sea level in adjacent and Mediterranean-type seas with the corresponding height in the oceans.

Before proceeding to a more detailed account of the present knowledge of the problem of a world-wide mean sea level the results given by Montgomery (1969) must be quoted. Montgomery paid special attention to the choice of reference level for the determination of the dynamic topography of the physical water surface. According to Montgomery, the optimum reference level over the world oceans seems to be 2,000 decibar or slightly more. The currents in the oceans tend to weaken with increasing depth down to at least 2,000 m, and the reference level should therefore preferably exceed 1,000 decibar. On the other hand, currents are observed in some oceanic regions to increase in strength in the vicinity of the bottom.

Lisitzin (1965) computed the mean sea level in dyn. cm — correponding to the mean specific volume anomalies — relative to the 4,000-decibar surface for different latitudes between 60°N and 50°S and the average sea level for the separate oceans. The results are given in Table XXXII.

The table reveals that for all latitudes the mean sea level is lowest in the Atlantic Ocean, while the highest values — if latitude 20°S is left out of consideration — are to be found in the Pacific Ocean. In total, the Pacific Ocean stands 72 dyn. cm higher than the Atlantic Ocean, but only 36 dyn. cm higher than the Indian Ocean. Reid (1961b) estimated that with respect to the 4,000-decibar surface the mean sea level in the Pacific Ocean is about 68 cm higher than in the Atlantic Ocean, a result which is in close agreement with the values in Table XXXII.

Table XXXII indicates, in addition, the marked latitudinal distribution of mean sea

TABLE XXXII

MEAN SEA LEVEL IN DYN. CM RELATIVE TO THE 4,000-DECIBAR SURFACE FOR DIFFERENT LATITUDES OF THE OCEANS

Latitude	Atlantic Ocean	Indian Ocean	Pacific Ocean
60°N	170	–	240
50°N	215	–	254
40°N	251	–	297
30°N	277	–	358
20°N	277	323	359
10°N	273	318	328
0°	272	308	334
10°S	265	314	336
20°S	271	352	346
30°S	277	312	325
40°S	251	277	290
50°S	189	209	254
60°S	120	137	173
60°N – 60°S	243	279	315

level in the oceans. The lowest values occur in the higher latitudes of the two hemispheres and there is a secondary minimum in the zones around the equator. This secondary minimum is reached in the Pacific Ocean at the latitude of 10°N, in the Indian Ocean at the equator, and in the Atlantic Ocean at the latitude of 10°S. The maximum mean sea level occurs in all oceans and in both hemispheres at the latitudes of approximately 20° and 30°. From this highest position the mean sea level slopes steeply towards the poles.

In order to obtain a comparison of the data giving the relative mean sea level, or the specific volume anomalies, obtained by Lisitzin with some other results the data are compiled with the values for the potential specific volume anomalies determined by Montgomery (1958) and based for the Pacific Ocean on the results given by Cochrane (1958), for the Indian Ocean on those by Pollak (1958) and for the Atlantic Ocean on those by Montgomery himself. All the data concerned are reproduced in Table XXXIII. The deviation between the results for the whole world ocean amounts to 16 cl/ton. The mean specific volume in situ of the water column for the upper 4,000 m may be estimated to be approximately 14 cl/ton greater than the potential specific volume. The deviation is thus about 2 cl/ton and it may be ascribed, at least to some extent, to the use of different data. The seasonal cycle in water density is assuredly an important factor in this connection. It must also be taken into account that Montgomery and his colleagues based their computations on the total water-covered area, thus including the adjacent and Mediterranean seas, and also upon the water masses between the depth of 4,000 m and the bottom, while Lisitzin's values refer only to the deep-sea regions of the oceans from the upper layers to the depth of 4,000 m, covering the latitudes between 60°N and 60°S.

Some additional features characteristic of the regional distribution of the mean sea level in the oceans may be pointed out. The average height of the sea level in the northern hemisphere is 307 dyn. cm, while in the southern hemisphere it is 282 dyn. cm. The mean sea level is thus 25 dyn. cm higher in the former hemisphere than in the latter. This deviation may, according to Reid (1961b), be ascribed to the fact that the waters in the South Pacific Ocean are richer in salt than those in the North Pacific, while the North Atlantic Ocean is warmer than the South Atlantic Ocean.

Turning to a more detailed, even if in many respects schematic, picture of the distribution of the mean sea level in the deep-sea areas of the world oceans — mainly according to the results of Lisitzin (1965) — the chart reproduced in Fig. 30 may serve as an introduction. In addition to the anomalies of specific volume only the static effect of atmospheric

TABLE XXXIII

THE MEAN SPECIFIC VOLUME ANOMALIES COMPUTED BY LISITZIN AND THE MEAN POTENTIAL SPECIFIC VOLUME ANOMALIES ACCORDING TO MONTGOMERY (CL/TON)

	Pacific Ocean	Indian Ocean	Atlantic Ocean	World Ocean
Lisitzin	79	70	61	72
Montgomery	62	56	45	56
Differences	17	14	16	16

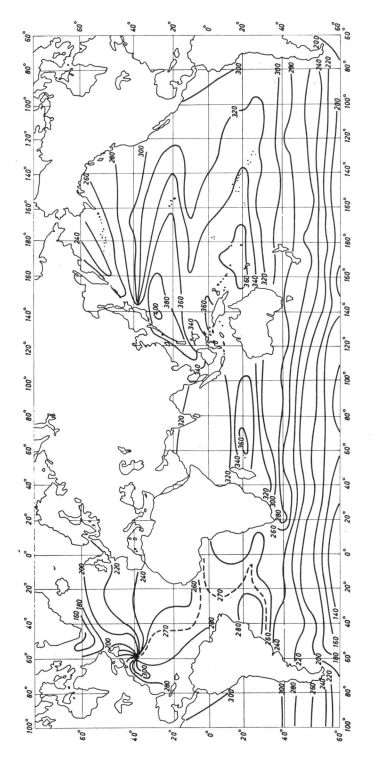

Fig. 30. The distribution of different heights of mean sea level (in dyn. cm) in the world oceans (Lisitzin, 1965).

pressure was considered as a correcting factor. This effect was investigated on a world-wide scale by Lisitzin (1961). The results obtained were more closely described in connection with the interaction between atmospheric pressure and sea level (pp. 61–62). On the contrary, the effect of the wind and the disturbing contribution of the anomalies of gravitational force and the constant deforming force of Moon and Sun were left out of account.

The chart given in Fig. 30 must be considered as a first, preliminary effort to reconstruct the deviations in mean sea level in the oceans. The distance of 20 dyn. cm between the isopleths is therefore quite reasonable. In order to illustrate the characteristics of mean sea-level distribution in the equatorial zone the isopleth representing 270 dyn. cm has been inserted in the central parts of the Atlantic Ocean. The isopleths in the chart do not reach the coasts, which is an indication that the particular perturbating effects on sea level in the coastal regions are completely disregarded in the chart.

Fig. 30 shows that the lowest mean sea level in the oceans, limited by the latitudes 60°N and 60°S, occur in the southern part of the Atlantic Ocean off the Antarctic coast, being only 130 dyn. cm. The highest mean sea level occurs, in complete correspondence with the results of Stommel (1964, 1965) in the western parts of all oceans. In the Atlantic Ocean the maximum height may be estimated at 300 dyn. cm, in the Indian Ocean at 360 dyn. cm and, finally, in the Pacific Ocean at 400 dyn. cm. According to these results fluctuations of the average height of sea level in the open deep-sea areas of the oceans situated between 60°N and 60°S have the approximate range of 270 dyn. cm.

The average sea level for all the deep-sea regions of the oceans considered in Fig. 30 is 289 dyn. cm. As will be shown in the following section, the mean sea level in the adjacent and Mediterranean seas is generally lower than that in the oceans. If, in addition, the highly pronounced decrease in mean sea level towards the poles is taken into account, there is no doubt that the average height of the whole area covered by water is somewhat lower. There is so far no possibility of determining the exact value of this height. As a first approximation a height of 280 dyn. cm may be ventured and this value will be made use of below.

THE ADJACENT AND MEDITERRANEAN SEAS AND THE TRANSITION AREAS BETWEEN THEM AND THE OCEANS

The adjacent and Mediterranean seas constitute roughly 11% of the total area covered by sea water. Nevertheless, only about a quarter of the area occupied by the seas has the character of deep water. There seems therefore to be no reason to divide the seas into deep and shallow regions, when it is more appropriate to investigate them as unities. Before commencing this task it may be of interest to mention a few figures which are typical of the thermo-haline structure in the seas. For the Mediterranean Sea there are available a number of oceanographical observations which cover the water column from the surface down to the depth of 4,000 decibar. The dynamic computations of these observed data have shown that the mean sea level varies between −98 and −115 dyn. cm and decreases slowly from the west to the east. These data indicate that the height difference in mean sea level, determined in a purely theoretical way and without paying attention to the actual conditions, can be estimated for the Atlantic Ocean west of the Strait of Gibraltar on the one hand, and for the eastern Mediterranean on the other hand at 350 dyn. cm.

This marked height difference can assuredly have no real background. It is self-evident that the deviation in the height of the water surface of two relatively deep sea basins separated by a shallow and narrow transition area is dependent on the anomalies of the specific volume of the water columns on the two sides of the passage, while the density of the water filling the depressions below the depth of the threshold cannot contribute to the relative heights of sea level. The computations of the slope of the water surface in narrow transition areas must therefore always be based on knowledge of the physical properties of the upper layers above the sill depth, or, since strong in- and outgoing currents generally occur in the passages which are typical of the different water layers, to the average depth marking the boundary between these currents.

In this respect it may be appropriate to pay more attention to the average deviations in water density in two sea basins linked by a narrow passage, since these deviations are decisive for the general outlines of the water interchange between the basins. If water density is lower in the adjacent basin than in the ocean or the sea situated outside, an outgoing current is the characteristic feature in the upper layers, this current being compensated by a current which is directed inwards in the lower and bottom layers of the transition area. The best-known examples of this type of water circulation in connecting regions are provided by the Baltic Sea and the Black Sea. On the contrary, if water has a higher density in the adjacent basin than in the outer sea or the ocean, the upper current runs inwards, and the compensating lower current has the opposite direction. This type of water circulation is characteristic of the Strait of Gibraltar between the Mediterranean and the Atlantic Ocean. A similar pattern of water exchange is also shown in the Strait of Bab-el-Mandeb between the Red Sea and the Gulf of Aden in the Indian Ocean.

The Mediterranean Sea, with the specific volume of the water markedly deviating from that in the eastern parts of the Atlantic Ocean, presents an important and interesting study of the slope of the water surface in the Strait of Gibraltar. The stratification of the water and the currents typical of the transition area have to be considered first. It has already been mentioned above that the current in the upper layers of the passage is going east and in the lower layer west. Mainly as the consequence of the pronounced evaporation in the Mediterranean during the summer months, these currents show considerable seasonal fluctuations in strength. It is obvious that the upper current must increase in volume in order to counterbalance the decrease in the water volume brought about by evaporation. Defant (1961) gave the sill depth of the Strait of Gibraltar as 333 m. The boundary layer between the in- and outgoing currents is situated at a depth of approximately 100 m in winter, while it runs about 100 m lower during the summer months. A depth of 150 m may therefore be assumed to correspond to the average depth of the boundary. Defant determined that the mean down-slope of the physical water surface from the Atlantic Ocean to the Mediterranean is 0.6 cm per 100 km. Defant based his computations on the assumption that the depth of the current boundary is 142 m. The theoretical height difference of 0.6 cm per 100 km seems to be too slight to be real. New efforts have therefore been made to determine the height difference on the basis of observed data. This task is by no means an easy one. However, making use of the oceanographic data collected in the area concerned by the research vessels 'Dana', 'Atlantic' and 'Xauen', the conclusion can be drawn that a surface slope of 5 dyn. cm per 100 m seems to be the most acceptable in the transition and border regions between the Atlantic Ocean and the Mediterranean. Proceeding eastwards in the deep parts of the latter sea basin, the oceanographic data indicate a continuous decrease in the mean sea level. For instance, at longitude $5°E$, the decrease may be estimated at 25 dyn. cm. The decrease continues, if in a less pronounced way still further east. A rough estimate between $5°E$ and $20°E$ is 15 dyn. cm. The easternmost part of the Mediterranean seems to be characterized by a rather slow increase of the mean sea level. An estimate gives an approximate value of 10 dyn. cm.

Passing over from the open sea to the coastal regions, the disturbing effect of different meteorological elements, in particular that of the wind, must be taken into account. The predominant wind direction over the Mediterranean Sea is from the northwest with a pronounced stability during the summer months. Therefore, there seems to be hardly any doubt that the main sea level must be lower along the northern coast of the Mediterranean than along the southern coast. The isopleths running through the central parts of the sea basin must therefore have an orientation in the direction west–east. Fig. 31, which gives a schematic picture of the results outlined above, makes it also possible to estimate the mean sea level along the European coast of the sea. Finally, this estimate allows us to compare the results determined on the basis of oceanographic data with the values for mean sea level at the tide-gauge stations computed with the help of geodetic precise levelling. The Committee for the 'United European Levelling Net' (UELN, 1959) determined the relative mean sea level for a considerable number of these stations situated along the coasts of western Europe. In order to obtain a comparison between values for mean sea level achieved by means of different approaches to the problem, Table XXXIV has been compiled. In this table are given data estimated on an oceanographic basis with the help of the chart in Fig. 31, results based on precise levelling by UELN and values for mean sea level computed by Rossiter (1967) for isobaric conditions with the use of the method of 'oceanographic' levelling. The data refer to two stations on the Atlantic coast and four stations on the Mediterranean coast of Western Europe.

The deviations (A–B) and (A–C) in Table XXXIV indicate that the mean sea-level heights determined by different methods are fairly consistent. The results are more favourable than could be expected, since the oceanographic estimate could not be made according to Fig. 31 with a greater accuracy than 5 cm. In this respect it should be noted that deviations between cm and dyn. cm, amounting to 2% of the numerical values, are not taken into account in the estimated data.

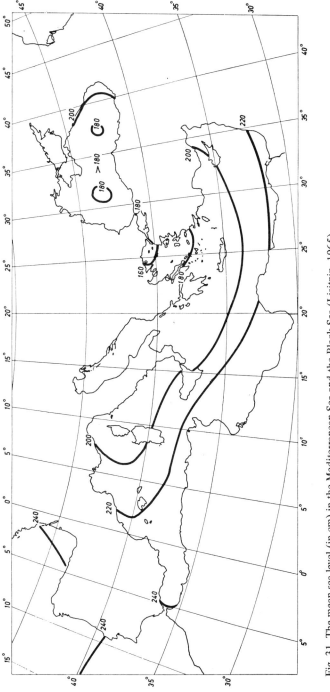

Fig. 31. The mean sea level (in cm) in the Mediterranean Sea and the Black Sea (Lisitzin, 1965).

TABLE XXXIV

THE MEAN SEA LEVEL (IN CM) AT 6 TIDE GAUGES ON THE ATLANTIC AND MEDITERRA-
NEAN COASTS DETERMINED BY DIFFERENT METHODS

Method	Brest	Lisbon	Alicante	Marseilles	Genoa	Trieste
A. Oceanographic estimation	225	242	230	210	195	190
B. Precise levelling	− 1.5	14.3	− 3.3	−17.3	−33.6	−33.7
A − B	227	228	233	227	229	224
C. Oceanographic levelling	3	−	−	−12	−	−28
A − C	222	−	−	222		218

Levallois and Maillard (1970) gave some results of the renewed first-order levelling in France performed in 1969. According to the authors, these results show with satisfactory likelihood the following features:

(1) a constant mean sea level along the coast of the Mediterranean from Malaga to Trieste;

(2) a difference in mean sea level of approximately 30 cm between the Atlantic Ocean and the Mediterranean Sea, the changes in sea level being closely concentrated around the Strait·of Gibraltar. The difference in sea level between Cadiz and Malaga is 15 cm.

The first-mentioned of the above results is rather strongly contradicted by older data and requires additional investigations based on oceanographic levelling. The results for the Strait of Gibraltar are fairly convincing. A height difference in sea level of 30 cm between Brest and Marseilles does not agree with the older results. The problem as a whole is thus so far unsolved.

Interesting details on the conformity between earlier oceanographic results for mean sea level and those achieved by means of geodetic precise levelling were given in a more extensive study on this subject by Lacombe (1959).

Among the additional conclusions which may be drawn on the basis of Fig. 31, mention may be made of the mean sea level in the eastern Mediterranean at the approaches to the Dardanelles and at the northern entrance to the Suez Canal. In the former case the mean sea level is approximately 155 cm, in the latter roughly 230 cm. These results will be referred to in more detail in connection with the problem of the mean sea level in the Black Sea on the one hand, and in the Red Sea on the other hand.

The Black Sea is separated from the northeastern part of the Mediterranean, represented by the Aegean Sea, by the Bosphorus, the Sea of Marmara and the Dardanelles. The two straits are fairly shallow, the sill depth of the Bosphorus amounting to 37 m and that of the Dardanelles to 57 m. The stratification of the water layers and the currents in the straits are well-established as the result of comprehensive studies performed by Merz and Möller (1928). The current in the upper layer in the two straits and the Sea of Marmara is directed from the Black Sea to the Aegean Sea and in the lower layers the current moves in the opposite direction. The boundary depth between the two currents in the Bosphorus may be estimated at about 30 m, sloping, however, sharply downward towards the northeast. In the Dardanelles the boundary layer is situated at a depth of approximately 20 m. These general characteristics, which are completed by other data given by Merz and Möller, have allowed the estimate that there is an increase in mean sea level from the Aegean Sea to the Sea of Marmara entrance of the Dardanelles of approximately 10 cm, while in the Bosphorus the increase in mean sea level may be estimated at 25 cm. Starting from the height of 155 cm for the mean sea level at the entrance to the Dardanelles in the Aegean Sea, the corresponding level at the approaches to the Bosphorus in the Black Sea may be estimated at 190 cm. Neumann (1942) was able to complete the picture, showing that the Black Sea has two regions of minimum mean sea level, situated in the central parts of the basin. Within these regions, located to the southwest and the southeast of the Crimean Peninsula and characterized by two cyclonic current gyres, the mean sea level may be estimated at 180 cm. According to the results given by Neumann, it could also be estimated that the highest mean sea level in the Black Sea occurs off the Caucasian coast, being there slightly more than 205 cm. The estimate for the coasts of the Crimean Peninsula is 190 cm.

In spite of the fact that the Red Sea is adjacent to the Indian Ocean, it seemed appropriate to investigate this sea in connection with the Mediterranean, since it confirms, at least to some degree, the results obtained for the latter sea. It has already been mentioned above that the mean sea level in the Mediterranean at the northern entrance to the Suez Canal may be roughly estimated at 230 cm. The results for geodetic precise levelling along the canal performed in the years 1923–1925 showed that the sea level stands 24 cm higher at Suez than at Port Said (Rouch, 1948). It must, however, be pointed out that the difference in mean sea level between the two ends of the canal has a pronounced seasonal variation. The sea level at Suez is higher than at Port Said during the greater part of the year except in the late summer months, when the conditions are reversed. The maximum average difference in sea level of Suez (Port Taufiq) over Port Said occurs in April, amounting to 36.3 cm (period 1924–1937) and 29.2 cm (period 1956–1963). The maximum average of Port Said over Suez is observed in September, varying between 2.3 cm for the former period and 10.7 cm for the latter period (Morcos, 1970). On the basis of the height differences Morcos (1960) estimated the average departure in mean sea level between the two localities at 17.6 cm for the period 1924–1937, and Morcos and Gerges (1968) at 12.3 cm for the years 1956–1963. It is thus difficult to draw general conclusions. However, taking into account that the mean sea level stands considerably higher at Suez than at Port Said during a great part of the year an average value of 20 cm could possibly be considered a reasonable solution. The mean sea level at the northern entrance of the Gulf of Suez should then reach a height of approximately 250 cm. This height may also be determined by considering the conditions in the Indian Ocean, and taking into account the sill depth and the average direction of the current in the Strait of Bab-el-Mandeb and, in addition, the fluctuations in the density of the water in the Red Sea. Assuming the mean sea level in the Gulf of Aden to be 305 cm, taking the sill depth in the transition area at 185 m, the boundary layer between the inward-directed upper current and the outward-moving lower current at approximately 100 m, and utilizing the oceanographic data collected during the expeditions of the 'Atlantis', the average sea-level height at the southern entrance to the Red Sea can be estimated at 295 cm. The decrease in mean sea level in the Strait of Bab-el-Mandeb is thus roughly 10 cm. Within the Red Sea a continuous decrease in mean sea level towards the northwest may be noted, and an estimate over the entire basin is 30 cm. The decrease brings about the height of mean sea level of 265 cm at the southern entrance to the Gulf of Suez, a value which is 15 cm higher than that computed for the northern extremity of the gulf. This descrepancy, which is not very pronounced and which may at least to some degree be compensated by a decrease in mean sea level from south to north within the Gulf of Suez, gives on the one hand a concept of the reliability of the estimated values for mean sea level, and indicates on the other hand that the applicability of the results of dynamic computations for the deep-sea parts of the oceans is rather restricted when considering coastal regions, if no allowance is made for the contribution of the meteorological elements. However, the prevailing wind direction over the Red Sea is from the northwest and results in a piling-up of the water not only in the southern parts of the Gulf of Suez but also in those of the Red Sea. The approximate deviation of 15 cm may therefore be easily accounted for by the disturbing meterological effect and the inaccuracy of the dynamic estimates. A schematic reproduction of the above results for the Red Sea is given in the chart in Fig. 32.

The Baltic Sea is assuredly one of the most studied sea regions as far as the mean sea level and its variations are concerned. The problem of the mean sea level has been discussed in greater detail for the Baltic Sea than for any other adjacent sea, mainly due to the fact that a pronounced land uplift is characteristic of the northern and central parts of this region. This contributing factor renders the whole problem not only more interesting but also more significant than in other coastal areas.

According to a rather rough estimate given by Bowden (1960), the effect of water density in the North Sea, the transition area around Denmark and the Baltic Sea brings about an increase of 18 cm in mean sea level after passing through the Danish Straits. This estimate is, however, very approximate owing to the complicated distribution of salinity, its highly pronounced variation in time and the complex topography of the sea bottom. In the main basin of the Baltic Sea the increase in mean sea level is according to Bowden 11 cm.

Lisitzin (1962) determined the increase in mean sea level in the Baltic proper as a consequence of the density effect at 7.9 cm. A further rise in mean sea level caused by decreasing water density, amounting in the Gulf of Bothnia to 4.7 cm and in the Gulf of Finland to 3.7 cm, may be noted (Lisitzin, 1957b, 1958a). Since for the Baltic proper the value computed by Lisitzin is about 30% lower than the value given by Bowden, it seems reasonable to make a corresponding reduction in Bowden's estimate for the Danish Straits, arriving at the height difference in mean sea level of 12.5 cm for the two entrances to this transition area. The added density effect in the particular basins results thus in a mean sea level in the innermost parts of the Gulf of Bothnia and the Gulf of

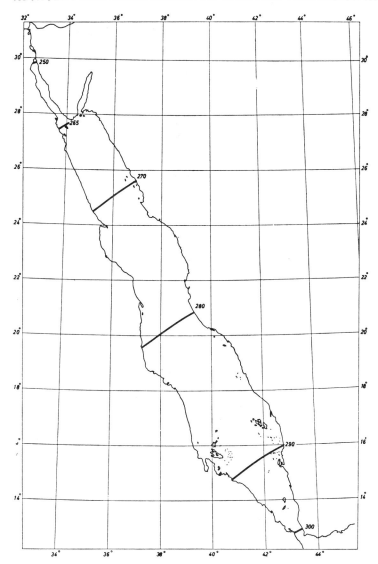

Fig. 32. The mean sea level (in cm) in the Red Sea (Lisitzin, 1965).

Finland, standing respectively 25 cm and 24 cm higher than at the North Sea entrance to the transition area. In the Baltic proper and its extensive gulfs the contribution of wind and atmospheric pressure very markedly reinforces the density effect. This contribution was determined by Lisitzin for the Baltic proper and the Gulf of Bothnia at 10.9 cm, while for the Baltic proper and the Gulf of Finland the corresponding value was computed at 7.4 cm. The final height difference for the former two basins is thus 36 cm, for the latter two basins 32 cm. Compared with the dynamically determined sea-level heights in the world oceans, the mean sea level at Kemi in the northernmost parts of the Gulf of Bothnia may be estimated at 256 cm, and at Hamina in the eastern part of the Gulf of Finland at 252 cm.

The above data, although only approximate, can be confirmed by the results achieved by geodetic precise levelling and presented by the Committee for the UELN (1959). These measurements showed that the average height difference in mean sea level between Kemi and Smögen — the latter station being situated half-way between the Norwegian–Swedish border and Gothenburg — is 36 cm and in close agreement with the above results. Finally, mention must be made of the determination of mean sea-level heights by Rossiter (1967). In an extensive study on the annual sea-level variations in European waters Rossiter computed the mean sea level for a considerable number of stations, correcting the relevant data for the influence of meteorological elements according to a method indicated by Doodson (1960). In addition, the perturbating effect of the nodal tide was taken into account by Rossiter, while the contribution of the spatial distribution of water density was left out of consideration. In this way Rossiter computed a mean sea-level difference between Kemi and Smögen of 25 cm. This value implies that the contribution of water density should be less pronounced than has been assumed above. The problem thus requires additional investigations, although the general trend is already fairly distinct.

The chart in Fig. 33 reproduces graphically the distribution of mean sea-level for every 5 cm in the Baltic Sea and the transition area around Denmark.

Starting from the results for the Gulf of Finland as given in Fig. 33, and taking into account the pronounced decrease in water density towards the easternmost part of the gulf and the marked effect of wind-caused piling-up of the water in the shallow and narrow Bay of Kronstadt, in can be estimated that the mean sea level at Leningrad reaches 260–270 cm. Unfortunately, no more recent data on precise levelling performed between the Baltic and the Black Sea are available, but according to the results of an older precise levelling the former sea stands 88 cm higher than the latter (Rouch, 1948). The most logical assumption in this connection is that the mean sea level for the Baltic refers to Leningrad, while the Black Sea is represented either by Sevastopol on the Crimean Peninsula or Odessa on the northern coast of the sea. For both these latter localities the mean sea level may be supposed to be approximately 190 cm. The height difference between the two seas should then amount to 70–80 cm. The departure from the results for precise levelling may be easily accounted for by the approximations used in determining the mean sea level in the concerned basins and also by a possible inaccuracy in the old geodetic levelling over a distance considerably greater than 1,500 km.

With the exception of the Red Sea, outside Europe sufficient oceanographic, sea-level and geodetic data to make possible a comparative determination of relative heights of mean sea level for long distances along the continental coasts are only available for the United States. Considering firstly the 'American Mediterranean' it must be mentioned that this sea consists of three main basins: the Caribbean Sea, the Yucatan basin (Bay of Campeche) and the Gulf of Mexico. The principal characteristics of this sea are markedly

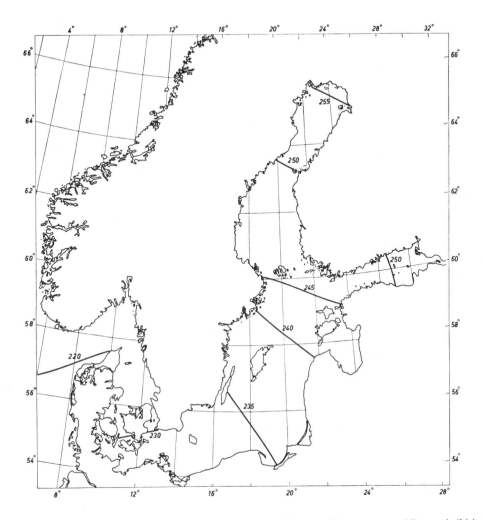

Fig. 33. The mean sea level (in cm) in the Baltic Sea and the transition area around Denmark (Lisitzin, 1965).

different from those of the basins investigated above, largely owing to the fact that its connection with the ocean is not restricted to a narrow and comparatively shallow channel, but consists of several passages of different breadth and depth between the Lesser and Greater Antilles and the mainland. In addition, the general pattern of water circulation in the American Mediterranean is not, as in the case for the seas described above, the consequence of gradient currents caused by differences in water density. The currents determining the water transport off the eastern coast of the United States are the continuation of the trade-wind driven currents and, farther to the north, the Gulf Stream.

These strong currents are the cause of a pronounced transversal slope of the water surface in some of the narrow parts of the 'American Mediterranean'. For instance, the average height difference in mean sea level between Cat Cay (Bahamas) and Miami (Florida) amounts, according to Iselin (1940), to 55 cm and according to Montgomery (1941a, b) to 52 cm, while Hela first (1952) determined this departure at 54 cm and later (1957) at 59 cm, and, finally, Wunsch et al. (1969) at 66 cm. The average value of these differences in mean sea-level height, 57 cm, is assuredly close enough to the actual conditions.

On the basis of oceanographic data collected by the 'Crawfort' expedition in 1958 it was possible to compute the anomalies of the dynamic heights in the Atlantic Ocean off the coasts of the Antilles on the one hand and in the Caribbean Sea on the other hand. There do not seem to be very pronounced deviations in mean sea-level heights, if the regions concerned are comparable in position in regard to the effect of the Coriolis parameter on the currents.

Precise levelling performed in the United States gives a considerable amount of mean sea-level data for tide-gauge stations situated along the coast of the Gulf of Mexico and the Atlantic Ocean. In this respect it must be especially pointed out that there is a marked difference between the older results of precise levelling published by Avers (1927) and those given by Braaten and McCombs (1963) as a consequence of an adjustment of a first-order levelling. According to the more recent results, the lowest mean sea level along the United States coast is noted at Key West (Florida). Progressing towards the north along the Atlantic coast of the United States the mean sea level increases continuously. The rise from Key West to Portland (Maine) amounts to 58 cm. The increase in mean sea level may be ascribed to the withdrawal of water by the Gulf Stream from the immediate vicinity of the coast and the occurrence of a counter-current along the coast. In this connection it may be appropriate to mention the results of Sturges (1968), although they may be considered to refer more closely to the next section. Sturges in his paper paid special attention to the topography of the sea surface in the area dominated by the Gulf Stream between Bermuda and the east coast of the United States. On the one hand, observations on ship drift were used and the effect of wind considered. The difference in sea-level height was found to be 110 cm. On the other hand, the deviations in sea-level heights across the Gulf Stream were computed by Sturges on the basis of oceanographic data and the estimates of deep-sea flow. This method resulted, after the application of some corrections, in 100 cm as the height difference in sea level relative to the 2,000-decibar depth. Starting from these results Sturges discussed the slope of the water surface along the Atlantic coast of the United States. The pronounced increase in mean sea level from south to north is in disagreement with the requirements of the observed flow of the Gulf Stream. Further measurements on land and in the ocean are thus indispensable in solving this problem.

The isopleths in the chart, Fig. 34, reproduce the general outlines given above for the American Mediterranean and the east coast of the United States. The increase in mean sea level from the south to the north along the eastern coast of the United States in this chart

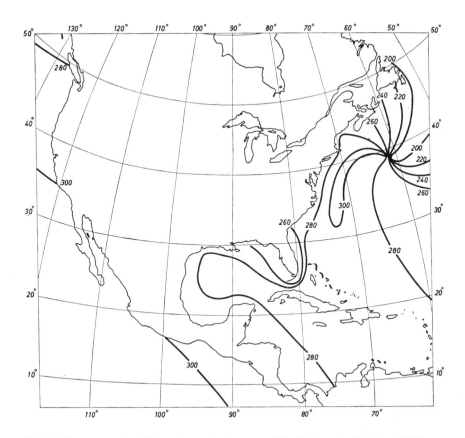

Fig. 34. The mean sea level (in cm) around the coasts of North America (Lisitzin, 1965).

is supposed to be about 35 cm, which is in agreement with the results of the earlier precise levellings published by Avers.

The Panama Canal forms a connecting link between the Pacific Ocean and the Caribbean Sea, and also between the Pacific and the Atlantic Oceans. At the northern entrance to the Panama Canal the mean sea level is approximately 275 cm while on the Pacific side it may be estimated at somewhat below 300 cm; let us say 295 cm. The height difference of 20 cm deviates by 10% from the results of precise levelling across the Isthmus of Panama, which according to Avers was 18 cm. Reid (1961b) gave more recent information on mean sea-level height difference across the Panama Canal zone. This information, which is based on a personal communication from the United States Coast and Geodetic Survey, states that 'mean sea level at Balboa on the Pacific coast is about 3/4 foot higher than the mean sea level on the Atlantic coast'. The height difference should thus amount to roughly 22 cm. This result, too, is in conformity with the estimate given above. The deviation from the older value may be easily accounted for by possible vertical movements of the

Earth's crust in the marshy region or, quite possibly, by the inaccuracy of the earlier measurements.

THE NEAR-SHORE REGIONS IN THE OCEANS AND SEAS

Through the Panama Canal, the Pacific Ocean and the western coast of the American continent have been reached. The adjustment of precise levelling in the United States (Braaten and McCombs, 1963) gives six values illustrating the mean sea level along the Pacific coast of the country. In order to give some numerical examples of the differences in mean sea level between the Pacific and the Atlantic coasts of the United States, it may be mentioned that the water surface at Neah Bay (Washington) stands, according to precise levelling, 71 cm higher than at Portland (Maine), while the height difference in mean sea level between San Francisco (California) and Norfolk (Virginia) is 62 cm and between San Diego (California) and Fernandina (Florida) 65 cm. These figures show that the Pacific Ocean stands, on average, approximately 66 cm higher than the Atlantic Ocean. Dynamic computations for the deep-sea areas of the two oceans have resulted in a mean sea-level departure of 72 cm (Table XXXII). The deviations between the two height differences may, at least to some extent, depend on the fact that the mean sea level is as a rule lower in the eastern parts of the oceans than in their western parts, the highest values being reached in the latter areas.

A closer study of the mean sea-level data based on precise levelling along the Pacific coast of the United States reveals a continuous increase from south to north. This increase amounts from San Diego to Neah Bay to 46 cm. For the tide-gauge stations situated to the north of San Francisco this increase is highly accentuated. However, according to Sverdrup et al. (1942), this result 'is based on records of sea level at the mouth of the Columbia River (Port Stewens), where the outflow of fresh water may account for the higher sea-level, and further to the north, on records of sea level at great distance from the open coast. The rise towards the north is therefore not well established'. Moreover, interesting results on the mean sea level and the slope of the water surface outside the Pacific coast of the United States have been given by Sturges (1967). Sturges' computations, based on oceanographic data relative to the 1,000-decibar surface, suggested that the considerable increase in mean sea level from San Diego to Neah Bay is non-existent. Conversely, mean sea level was found by Sturges to be at Neah Bay 9 cm lower than at San Diego. Sturges has, in addition, pointed out that according to a careful study the 1,000-decibar surface is sufficiently level as a reference surface along the coast. Moreover, the departure of 9 cm in mean sea-level height determined with the help of dynamic computations is consistent with the effect of the changes on latitude as the Californian current flows southward. The decrease in mean sea level determined by Sturges is also in better agreement with the general picture of the distribution of mean sea-level heights in the eastern parts of the Pacific Ocean, as presented in the chart in Fig. 30.

It has already been mentioned above (p. 150) that the average height of sea-level for the whole hydrosphere can be estimated at 280 cm. In spite of the difficulties in connecting the water surface heights in the deep sea with sea levels along the coast, a procedure that necessarily results in a rough estimate, a mean sea level corresponding to 280 cm may reasonably be ascribed to several coastal localities. It may be sufficient to mention one such example; Portland (Maine) in the United States ofAmerica. It must be kept in mind that Portland was the starting point for the American precise levelling system and that since its tide-gauge station has been in uninterrupted operation for a time-span exceeding half a century, it could possibly in the future, if the approximate value of 280 cm is proved to correspond with actual conditions, serve as a comparative starting point and reference surface for all measurements which require a knowledge of mean sea level.

Finally, it may be of interest to point out that there exists a height difference amounting to 52 cm between the mean sea level of 280 cm determined for the entire area covered by water and the zero level at Amsterdam used by the Committee for the United European Levelling Net (UELN) as the reference level for all geodetic determinations.

CHAPTER 6

LONG-TERM (SECULAR) CHANGES IN SEA LEVEL

VERTICAL MOVEMENTS OF THE EARTH'S CRUST

The vertical movements of the Earth's crust are, according to the theory advanced in 1865 by the Scottish geologist Thomas Jamieson, the consequence of ice masses which covered large parts of the Earth during the Glacial period and their subsequent disappearance. These ice masses exterted a tremendous pressure causing the Earth's crust to subside, but the crust started to re-occupy its original position as soon as the ice pressure began to decrease. This process has continued after the disappearance of the ice and is still taking place. In recent years Jamieson's theory has been criticized, in particular in geological and geodetic circles. In connection with sea-level investigations the causes of the vertical movements of the Earth's crust are, however, a question of secondary significance. It is the numerical rate of land uplift or land subsidence to which attention must first be paid.

The consequences of land uplift upon the mean sea level in Fennoscandia, i.e., the continuous decrease of the average water height, have been known for centuries and the first approximate value for this decrease in sea level for the entire Baltic Sea — amounting to 4.5 feet, or 135 cm, per 100 years — was given by Anders Celsius (1743) as early as the mid-eighteenth century. Sporadic observations on the decrease in sea level in the northern parts of the Baltic basin are still older. The interest in this phenomenon and its significant consequences resulted in the middle of the nineteenth century in the erection of the first tide poles, and some decades later in the construction of the first tide gauges.

The first approximate chart demonstrating the distribution of the rates of land uplift in the Baltic Sea was given by Sieger (1893). The work was later continued by a number of Finnish oceanographers. The investigations performed by the following scientists may be mentioned in this connection: Blomqvist and Renqvist (1914), Witting (1918, 1922, 1943), Hela (1953) and Lisitzin (1964a). If to this list are added the researches performed by Bergsten (1930) in Sweden, Gutenberg (1941) in the United States, Model (1950) in Germany and especially Rossiter (1960, 1967) in Great Britain, the great interest which has been shown in this phenomenon may easily be noted. At the present time a large amount of data on the secular variations in sea level along the coasts of the Baltic — in some of the cases mentioned above, however, referring only to the Finnish coasts — are at our disposal, which on the one hand illustrate the main features of the phenomenon, but reveal on the other hand considerable discrepancies in separate results for the individual stations.

In this connection it must, however, be pointed out that in order to achieve satisfactory results the determination of the secular changes in mean sea level must be based not only on comparatively prolonged series of observations or records, but also on data which are highly accurate. Moreover, it must always be kept in mind that the annual averages for sea-level data are affected by a great variety of disturbing factors. Most of these factors have already been discussed in more detail in the previous chapters, but it may be appropriate to summarize them here. The most important and easily surveyable of the contributing factors are:

(1) Meteorological contribution: atmospheric pressure, wind, precipitation and evaporation.

(2) Oceanographic contribution: water density and currents.

(3) Hydrological contribution: river discharge.

(4) Eustatic contribution: melting of continental ice in polar regions.

(5) Astronomical contribution: the nodal tide with the period of 18.61 years.

(6) Technical contribution: the effect of a possible vertical movement in the gauge foundation itself or of accidental errors in the operation of the apparatus or in the readings of the recorded curves. The former of these effects is, generally, eliminated from the records by precise levelling which should be made to the closest bench mark every year; the latter by continuous controlling measurements which should be taken as frequently as possible, preferably once a week, while errors in the readings may be corrected by comparing the results for neighbouring stations. Although some of these effects may not occur with modern instrumentation, the determination of the rate of land uplift and land subsidence must always be based on prolonged series of data and older results have thus to be taken into account.

Among the different perturbating factors the meteorological effect, especially that of the wind, involves the most marked contribution to sea-level fluctuations in shallow basins such as the Baltic. In order to achieve reliable data in connection with the determination of the rate of the secular variation, it is imperative that this effect is eliminated from sea-level records. This elimination may be carried out in many different ways, of which the two most important will be briefly described.

(1) The removal of the meteorological effect from the data may be performed by utilizing values for atmospheric pressure at stations located around, and in some cases outside, the sea basin and deducing a general law for the influence of atmospheric pressure and the corresponding winds upon the water surface. Witting (1918) was the first to choose this method, introducing in his computations the so-called anemo-baric law as the fundamental basis. Rossiter (1960, 1967) proceeded in an analogous way, using, however, in his investigations a quite different scheme for the elimination of the meteorological effect on the sea-level data. He based his results on the data on atmospheric pressure of three stations, preferably forming an equilateral triangle covering the sea region over which the wind-generating force may be expected to act. In addition he considered the effect of the nodal tide on the sea level. (See p. 206)

(2) The contribution of the meteorological and, in addition, of the oceanographic and

hydrological elements may be eliminated on the basis of a comparison of the sea-level data themselves. Model followed such a way of proceeding in his study, and Hela developed his own more detailed and logical scheme for such a method. Lisitzin has, with some slight modifications, followed the method outlined by Hela. The principal modification consisted in the fact that Lisitzin also paid attention to the nodal tide, eliminating in her computations the perturbating effect of this tide on the annual sea-level data.

It lies assuredly beyond the boundaries of this presentation to describe all these methods in more detail and reference is therefore made to the particular studies already mentioned. Some general remarks on the reliability of the two main types of methods may, nevertheless, be of interest. In all the cases where the final results were based on a general scheme for the distribution of atmospheric pressure, the local effects, which are doubtless fairly pronounced in some cases, could not be eliminated from the sea-level data. The disturbing effect of water density was not taken into consideration by Rossiter. It must also be remembered that the data on atmospheric pressure, although as a rule reliable, may in some cases be deficient and introduce a new source of inaccuracy into the whole mathematical system. By removing the disturbing factors from the data themselves the meteorological and oceanographic effects have both been taken into consideration, at least to some extent. As Hela pointed out in his study, meteorological and oceanographic factors affect the annual sea-level records in the Baltic Sea area mainly in two ways: by the variation of the total water volume in the Baltic basin, caused to a more-or-less marked degree by forces acting outside the basin and being of the same magnitude for the whole sea, and by the changes of the slope of the water surface, which have to be taken into account as a largely local or regional characteristic. This slope of the water surface is a feature typical of the Baltic Sea due to the prevailing distribution of the atmospheric pressure and wind over the basin and of the differences in water density and has already been discussed here (pp. 128–137, cf. Lisitzin, 1957b, 1958a, 1962). However, a certain disadvantage of the method given by Hela should not be overlooked here. This disadvantage consists in the fact that the final results are referred to a selected sea-level station. An inaccuracy in the rate of land uplift at this station will necessarily also be reflected in all other computed rates. Moreover, it is by no means excluded that possible long-term variations of the meteorological effects will be integrated in the rate of land uplift.

Furthermore, it must be kept in mind that the eustatic increase in sea level, as will be shown in more detail in the next section, can hardly be expected to be strictly continuous and uniform. On the contrary, pronounced departures from the general course are highly probable. It must also be remembered that the increase in air temperature is accompanied not only by the acceleration in the melting processes of the continental ice, but also by changes in water volume in the oceans. The final result is therefore not easily foreseen. Possible deviations from the average values will, however, be integrated in the computations as a part of the changes in the total water volume in the Baltic.

Independently of the method on which the computations are based, there are considerable difficulties in eliminating completely the different perturbating effects and because of this the selection of the research period may in some cases be of as much significance as the choice of the method itself. Of course, the length of the period is extremely important and as lengthy series of observations as possible should be considered in every particular case. It must, in addition, be kept in mind that the meteorological conditions characteristic of the beginning and the end of the selected period may largely affect the results. If the average sea level had been comparatively low during several years at the beginning of the period, as, for instance, in the early 1940's, and had the elimination of the perturbating influence not been sufficiently effective, the computed rate of the secular variations would turn out noticeably too low. In its most extreme form these difficulties were demonstrated by the rates of land uplift computed by Rossiter (1960). There is a difference of approximately 4 mm per year for the rates of land uplift computed from the crude sea-level data and those corrected for the meteorological effects.

In this connection arises the significant question of whether the rate of the long-term movements of the Earth's crust is continual or more or less spasmodic. Unfortunately this question cannot yet be solved by oceanographers alone on the basis of sea-level data. The solution requires quite new methods and, in addition, a close collaboration with geodesists and seismologists. In spite of the fact that there are some indications of a weak retardation of the rates of land uplift in Finland and that the results arrived at by different authors vary, it seems at the present time adequate to assume a practically constant course for the vertical movement of the Earth's crust. The main aim should therefore be the determination of the average rates, attributing deviations to the disturbing effects and in particular to their unsatisfactory removal from sea-level data, and possibly to some extent to inevitable errors inherent in the original records.

Table XXXV gives a comparison of the rates of land uplift computed by different authors with the help of sea-level records. The use of the term 'land uplift' is not quite appropriate in this case, since the rates actually correspond to the decrease in sea level which is the result of land uplift and the eustatic factor. In this connection it may be emphasized that the rates given by Lisitzin, for instance, coincide within the limits of mean probable deviations with those determined by Kääriäinen (1966) on the basis of geodetic precise levelling. A similar statement may also be made concerning the data computed by Lisitzin and the more recent series of rates given by Rossiter (1967). Table XXXV allows us once more to establish that the general features of the distribution of the rates along the Finnish coast is on the whole similar in all cases. The sole pronounced deviation consists in the fact that, according to the results obtained by Model and all later authors, the rate of land uplift is most pronounced at Pietarsaari, while from the data determined by Witting this feature is found in the northernmost part of the Gulf of Bothnia, i.e., in the region between Kemi and Raahe. A graphical presentation of the data computed by Lisitzin is given in the chart in Fig. 35.

There are some interesting details which may be pointed out concerning the mean deviations given in Table XXXV. For the earlier series of rates, determined by Rossiter, the mean deviations are remarkably uniform, while the more recent data from Rossiter show very distinctly that the mean deviations decrease as soon as the length of the period on which the computations are based increases. For instance, the mean deviation for

TABLE XXXV

THE SECULAR VARIATION IN SEA LEVEL ALONG THE FINNISH COAST IN MM PER YEAR FOR DIFFERENT PERIODS AND ACCORDING TO DIFFERENT AUTHORS *

	Witting (1918) 1898–1912	Witting (1943) 1898–1927	Model (1950) 1904–1937	Hela (1953) 1922–1951	Rossiter (1960) 1940–1958	Listizin (1964) 1924–1960	Rossiter (1967) all available data
Kemi	–	(11.7)	7.2	6.4 ± 1.2	8.5 ± 1.4	7.3 ± 0.9	6.7 ± 0.5
Oulu	10.3	10.7	6.3	6.3 ± 1.2	7.0 ± 1.3	7.1 ± 1.0	6.1 ± 0.3
Raahe	–	(11.7)	7.5	7.4 ± 0.9	8.0 ± 1.4	7.8 ± 0.7	7.1 ± 0.4
Pietarsaari	–	–	8.7	7.6 ± 0.4	9.2 ± 1.4	8.2 ± 0.7	7.8 ± 0.4
Vaasa	9.2	8.7	8.0	7.2 ± 0.7	7.6 ± 1.3	8.0 ± 1.1	7.5 ± 0.2
Kaskinen	–	–	6.8	7.6 ± 0.7	6.7 ± 1.4	7.4 ± 0.7	7.5 ± 0.8
Mäntyluoto	7.4	6.6	6.8	6.5 ± 0.4	5.3 ± 1.4	6.4 ± 0.6	6.3 ± 0.3
Rauma	–	–	5.0	5.9 ± 0.4	3.9 ± 1.5	–	6.0 ± 0.6
Turku	–	(5.4)	3.3	4.8 ± 0.6	2.5 ± 1.5	4.4 ± 1.2	3.7 ± 0.5
Degerby	–	–	3.3	5.1	–	4.6	4.1 ± 0.5
Hangö (Hanko)	4.5	4.0	3.6	3.5 ± 0.6	1.1 ± 1.5	3.1 ± 0.5	3.0 ± 0.1
Helsinki	0.8	2.8	2.8	3.1 ± 0.4	–0.4 ± 1.5	2.5 ± 0.5	3.2 ± 0.2
Hamina	–	–	3.6	2.2 ± 1.0	–0.4 ± 1.5	–	1.8 ± 0.7

* Eustatic increase in sea level is not taken into account. Although the data are given as positive, they imply a decrease in mean sea level.

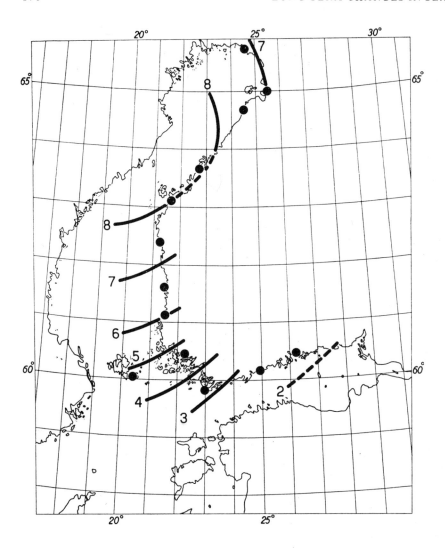

Fig. 35. The rate of the secular variation in mean sea level (in mm/year) due to land uplift along the Finnish coast (Lisitzin, 1964a).

Hanko (Hangö) amounts to 0.1 mm and for Helsinki to 0.2 mm. At the former station the records cover a time-span of more than 90 years and at the latter station the number of years of sea-level data exceeds 80. The most accentuated mean deviation – 0.8 mm – is noted at Kaskinen, where only 36 years of data are available. The values determined by Lisitzin show a quite different feature in the distribution of the mean deviations. Since all results were based on the same period the distribution of deviations indicates the local influence. The three stations Oulu, Vaasa and Turku, situated in the inner parts of the skerries and relatively far from the open sea, are characterized by the highest values of the mean deviations. The above remarks may indicate the reliability of the different methods on which the computations are based.

The Baltic Sea, including its extensive gulfs and in particular the Finnish coastal area, was chosen as the starting point for the determination of the contribution of land uplift to the mean sea level. This seemed to be the most natural way of proceeding, since the phenomenon is highly pronounced in this region, the available data fairly extensive and the number of different studies on the subject considerable. It is at the western coast of the Gulf of Bothnia, i.e., in Sweden, that land uplift reaches its highest values within the whole basin of the Baltic. Rossiter (1967), for instance, gave a rate of 8.9 mm per year for land uplift at Furuögrund. From this maximum value the speed of the vertical movement of the Earth's crust decreases relatively rapidly along the Swedish coast towards the south, according to Rossiter, being only 0.3 mm per year at Kungsholmsfort (Karlskrona) and changing to land subsidence with a rate of −0.8 mm per year for Ystad on the southernmost coast of Sweden.

The situation in Denmark seems to be fairly complicated. A survey for four Danish sea-level stations has been made by several oceanographers and their results completed by geodetic data are reproduced in Table XXXVI. The results of precise levelling are indicated relative to Hornbaek, where the vertical movement of the Earth's crust seems to be very weak. Since the rates computed by Rossiter (1967) for Gedser and Copenhagen did not agree with the results for geodetic levelling, at least not as well as might be expected from the estimates of accuracy, it seemed to be necessary to examine the reliability of the recorded data more closely. This study was performed by Borre (1970). Since a strong correlation could be established between the surface current at the light vessel 'Drogden' and the difference in mean sea level at Gedser and Copenhagen, this was utilized for the determination of sea level at these stations when the surface current was nil at Drogden.

Thereafter the method of Jakubovsky (1966) was applied in order to eliminate the influence of the main meteorological effects. After this correction the data were utilized to determine the secular variations. Unfortunately, the data in Table XXXVI show that

TABLE XXXVI

THE SECULAR VARIATION IN MEAN SEA LEVEL IN MM PER YEAR FOR SOME DANISH STATIONS ACCORDING TO DIFFERENT AUTHORS AND METHODS [*]

Station	Egedal (1955)	Rossiter (1967)	Borre (1970)	Precise levelling
Hornbaek	−0.02	−0.03 ± 0.17	−0.04 ± 0.47	0 ± 0.0
Korsor	0.72	0.78 ± 0.16	0.50 ± 0.33	0.74 ± 0.4
Gedser	0.91	1.04 ± 0.21	0.75 ± 0.36	0.23 ± 0.5
Copenhagen	0.24	0.23 ± 0.16	0.29 ± 0.43	−0.44 ± 0.3

[*] The eustatic increase in sea level is not taken into account. The negative data in the table imply a decrease in mean sea level.

these computations were not able to give the final answer to the question. Since the configuration of the Danish straits is rather complicated, it is difficult to build up a model corresponding to the natural conditions. In addition, the effect of water density should be taken into account before definite conclusions can be drawn. Finally, mention may be made of the rates of land subsidence for the Danish stations computed recently by Thomsen and Hansen (1970). These rates correspond closely to those given by Rossiter. Thomsen and Hansen established also that the rates for the vertical movement of the Earth's crust are less dependent on the method of revising the original sea-level data than on the time-span selected as the basis for the computations.

There is one more Danish sea-level station, Esbjerg, which has been studied more closely with regard to the long-term variation. The results giving the increase in mean sea level are the following:

	mm per year
Dietrich (1954)	1.14 ± 0.28
Egedal (1954)	1.29
Jessen (1955)	1.33 ± 0.19
Rossiter (1967)	1.48 ± 0.15

The rate of land subsidence computed by Dietrich was based on the sea-level records for the· period 1890–1950. All original data were carefully corrected for the effect of atmospheric pressure, the force and direction of the wind and the density of water. Egedal's result refers to observations from 1889 to 1954. Jessen used the observed data on sea·level for the years 1889–1961. These data have not been directly corrected for the influence of the different disturbing factors, but the annual mean sea-level values were computed from weighted monthly averages; the weights having been chosen in such a way as to moderate the effect of meteorological disturbances. Rossiter based his results on the period 1889–1962 and, as has already been mentioned above, eliminated from the data the meteorological effects and the contribution of the nodal tide. Once more the conclusion may be drawn that the rate of secular variation in mean sea level is only slightly dependent on the method used for the computations if the periods do not differ too markedly. The average value for the four rates given above is 1.31 mm per year and this value is practically within the limits of the mean deviations.

It is highly regrettable that the eastern and southern coasts of the Baltic Sea have not been studied as thoroughly as the northern and western coasts. This fact may be ascribed to two different circumstances. Firstly, it must be kept in mind that sea level records are generally not available from the first-mentioned regions, and secondly, that the movement of the Earth's crust is here rather weakly pronounced and so of less interest. A relatively accurate determination of the rate is also more difficult. However, it is possible to give a few data. Duvanin (1956) briefly mentioned that land uplift starts approximately at the line Leningrad–Riga–Copenhagen and increases in intensity towards the north. This statement is an indication that land subsidence occurs to the southeast of this line. This result has been confirmed by Pobedonoszev (1971, 1972). Rossiter (1967) gave 0.25 ± 0.39 mm per year as the value for land uplift in Riga. For the remaining parts of the coastal area, with the exception of Daugavgriva in the Latvian S.S.R., there is, according to Rossiter, a land subsidence with a rate varying between the limits of 0.5 and 1.0 mm per year but which does not show any marked regularity.

For the southern coast of the Baltic there are two series of data giving the rates of the long-term increase in mean sea level, i.e., in addition to the data computed by Rossiter the values given by Montag (1967). Montag considered in his study not only the contribution to the sea level of the atmospheric pressure, the piling-up effect of the wind and water density, at least as far as there are data available for these elements, but also possible and probably inevitable dislocations of the zero points of the tidal poles. A comparison of the results achieved by Rossiter and by Montag is given. The periods used by Montag are, as a rule, somewhat more prolonged than those on which Rossiter based his results:

	Rossiter	Montag
Baltijsk (Pillau)	0.80 ± 0.37	0.9 ± 0.2
Ustka (Stolpemünde)	–	0.8 ± 0.2
Swinoujscie (Swinemünde)	0.75 ± 0.15	1.1 ± 0.1
Arkona	0.07 ± 0.21	0.2 ± 0.4
Warnemünde	1.48 ± 0.21	1.4 ± 0.1
Wismar	1.49 ± 0.21	1.4 ± 0.2
Marienleuchte	0.98 ± 0.20	0.6 ± 0.2
Travemünde	2.34 ± 0.30	1.4 ± 0.2

A comparison of the two series shows an interesting trend in the subsidence data. From Arkona eastwards the values given by Montag are higher than those computed by Rossiter, while farther west the situation is the opposite. The most pronounced numerical difference occurs for Travemünde.

There is one more method available for determining the relative vertical movements of the Earth's crust between two localities. It is the method of oceanographic levelling which was developed by Jakubovsky (1966) to cover distances of several hundreds of kilometres. Taking into account the perturbating effect of atmospheric pressure, wind and air temperature – the latter as a substitute for water density – Jakubovsky was able to compute the relative vertical movements of the Earth's crust for the following pairs of sea-level stations: Degerby (Finland)–Ustka (Poland); Landsort (Sweden)–Hangö (Hanko, Finland); Kungsholmsfort (Sweden)–Baltijsk (Lithuanian S.S.R.). These results are compared with the rates of the relative vertical movements computed by different authors on the basis of sea-level records. The unit is 1 mm per year:

	Degerby–Ustka	Landsort–Hangö	Kungsholmsfort–Baltijsk
Jakubovsky	5.2 ± 0.5	−0.3 ± 0.3	2.1 ± 0.3
Lisitzin/Montag	5.2	–	–
Rossiter/Montag	4.9	–	–
Rossiter		0.0	1.1

With the exception of the connection Kungsholmsfort–Baltijsk the departures between the results for corresponding stations lie within the limits of the standard deviations.

In addition to the Baltic and the transition area around Denmark, Rossiter (1967) determined the rates of the secular variation in mean sea level for Norway, The Netherlands, the British Isles, France (only for Brest and Marseilles), Portugal (Cascais and Lagos) and Italy (Trieste). The rates computed for Norway show a fairly pronounced dispersion, indicating land uplift for some of the stations and land subsidence for others. For The Netherlands the rates vary between 1.3 and 3.0 mm per year, indicating very distinctly the well-known fact that the mean sea level is increasing. The increase in mean sea level seems also to be present for all the remaining stations whose data have been investigated by Rossiter. Nevertheless, it must be pointed out especially in this connection that the eustatic effect, which will be discussed in more detail in the next section, contributes to the increase in mean sea level, and that the results cannot be attributed exclusively to the vertical movements of the Earth's crust.

Thus it is highly probable that the rates for land subsidence determined by Pobedonoszev (1972) for the Black Sea and the Sea of Azov and for most of the concerned stations characterized by a speed varying between −0.6 and −1.6 mm per year are the consequence of the eustatic factor. On the contrary, in the White Sea there seems to occur, according to Pobedonoszev and Rosanov (1971), a land uplift along the coasts in the western part of the sea basin, while land subsidence is typical of the eastern coasts. For the Barents Sea the trend is less pronounced, especially on the northern coast of the Kola Peninsula. However, to the east of the Cape Canin Nos land subsidence seems to be evident, according to the results obtained by the two authors mentioned above.

It is only natural that studies concerned with the problem of the vertical movements of the Earth's crust are concentrated on certain coastal regions. These regions are, moreover, identical with those where sea-level records are numerous both in time and in space. In addition to the coasts of Europe this statement applies particularly to the coastal areas in the United States and Japan. Unfortunately, the methods used for the determination of the long-term variation are generally not sufficiently accurate to allow very satisfactory results. This concerns, for instance, the papers by Hicks and Shofnos (1965) and Hicks (1968, 1972b). The principal cause of this disadvantage is that the main purpose of these papers was technical. Hicks himself (1968) pointed out that the methods used 'should not be interpreted as an attempt to eliminate the meteorological and oceanographic contribution, but only to show changes in the trend within the large variations in yearly mean sea-level data.' Forty-three sea-level stations in the United States were analyzed by Hicks (1972b) for the period 1940–1970. It was established that for all the stations along the Atlantic and Gulf coasts of the United States, the increase in mean sea level already determined by Disney (1954) was confirmed, although the particular rates as given by the two oceanographers were highly deviating. Most of the curves reproduced by Hicks revealed a decreasing trend in mean sea level during the period from 1946 to 1964 in the northern part of the east-coast area, while more recent data indicated a return to rates characteristic of the years 1928–1946. Along the Pacific coast of the United States, with the exception of Alaska, there is also an increasing trend in mean sea level. In order to give numerical examples, the following data for the secular trend may be reproduced from Hicks (1972b):

Location	Trend 1940–1970, cm per year
Portland, Me	0.162
Boston, Mass.	0.107
New York, N.Y.	0.287
Baltimore, Md	0.259
Charleston, S.C.	0.180
Fernandina, Fla	0.125
Key West, Fla	0.073
Galveston, Texas	0.430
San Diego, Calif.	0.143
San Francisco, Calif.	0.192
Seattle, Wash.	0.259
Ketchikan, Alaska	0.030
Sitka, Alaska	−0.204
Juneau, Alaska	−1.305
Yakutat, Alaska	−0.503

Hicks (1973) later modified his results to some extent but the deviations, being for the most pronounced cases of the magnitude of 0.03 cm per year, are not large enough to change the general picture of the vertical movements of the Earth's crust along the coasts of the United States.

The rates for Alaska display a very pronounced land uplift, illustrating, according to Hicks (1968), the elastic or tectonic rebound from present localized deglaciation or the combination from present localized and general post-Wisconsin deglaciation. This emergence is centered in Bartlett Cove, Glacier Bay, where the rate of the maximum land uplift is approximately 4 cm per year. Ketchikan is located on the periphery of this area. The pronounced values of land uplift in the southeastern parts of Alaska were already established by Pierce (1960).

Hicks has in his investigations considered the eustatic increase of the water quantity in the oceans and seas. This effect was completely neglected by Roden (1963) in his research on sea-level fluctuations at Panama. Basing his study on the time-span from 1909 to 1962, Roden determined the total increase in mean sea level at approximately 8.5 cm for Balboa and at 7.1 cm for Cristobal. Roden ascribed this increase exclusively to land subsidence, while almost certainly a considerable part of these increases is a consequence of the eustatic effect. According to Hicks (1972b) there has been no secular trend at Cristobal during the period 1940–1970, the rate thus being zero.

The difficulties in determining the secular variation in mean sea level along the coasts of the United States in a more accurate way may be illustrated by the example of

San Francisco. For this station there are at our disposal four different estimates of the increase in mean sea level. Disney (1954) reached the conclusion that this increase amounts to 1.8 mm per year. This value was based on the data for the years 1898–1951 and no corrections were made for the contribution of the separate disturbing effects. Hicks (1968), applying a triangular weighting array to the yearly mean sea-level data, but no direct corrections for the meteorological and oceanographic parameters, obtained for the increase in mean sea level the value 1.5 mm per year for the period 1940–1966. For the period 1940–1970, Hicks (1972b), as already mentioned above, determined the positive rate of mean sea-level trend at 1.9 mm per year. Yamaguti (1962) took into account the fact that the mean sea-level curve for San Francisco from 1922 to 1958, after correction for water temperature, showed a more pronounced increasing trend than the uncorrected curve. He therefore drew the conclusion, that a slow subsidence of the Earth's crust relative to the mean sea level with a rate of approximately 4 mm per year, occurs, being thus more than twice as high as the rates given by Disney and Hicks. The possibility that the increase in mean sea level was brought about by the eustatic increase of the water volume in the North Pacific Ocean was excluded by Yamaguti, since the mean sea level at Aburatsubo, Japan, did not indicate a corresponding increase.

The vertical movements of the Earth's crust in Japan are also of considerable interest and extensive investigations have been performed, the work being facilitated by the great amount of observed sea-level data from this area. Tsumura (1963) determined the vertical movements of the Earth's crust for not less than 58 sea-level stations along the Japanese coasts. The results showed that the deviations between the separate stations were surprisingly pronounced. A few examples may be sufficient to illustrate his fact. At Osaka there is the extremely accentuated land subsidence at a rate of 58 mm per year, while the next highest rate of land subsidence, 23 mm per year, was determined for Tuba. Conversely, for the stations Abashiri and Maisaka a rate of land uplift amounting to somewhat more than 11 mm per year was given by Tsumura. The total number of stations characterized by land subsidence was 33; the number of those showing land emergence, 25. The average value for the vertical movements of the Earth's crust for all the Japanese stations, with the exception of Osaka where the marked subsidence of the ground is probably artificial, resulted in a land sinking of the rate of 1.14 ± 1.24 mm per year. The pronounced value of the standard deviation indicates, in correspondence with the considerable departures of the results for the separate stations, the inaccuracy of the numerical determinations. There does not seem to be any pronounced regional distribution of areas with land subsidence or land emergence. The highly seismic ground in the surroundings of the Japanese islands may easily account for this fact. Attention should also be paid to the fact that the period of sea-level observations on which the results of Tsumura's computations were based was relatively short, covering only the 10-year period 1951–1960.

Before leaving the problem of the vertical movements of the Earth's crust and their effect on the mean sea-level, a few words must be devoted to the great danger which at the present time is more and more threatening the historically and artistically world-famous city of Venice, situated in a lagoon in the northern part of the Adriatic Sea. This

danger consists mainly in the fact that an artificial land subsidence which is very pronounced, occurs. The rate of this subsidence is not quite uniform in different parts of the
city. According to Polli (1962a), the average increase in mean sea level in Venice, including not only the vertical crustal movement but also the eustatic factor, may be estimated
at approximately 3 mm per year. Frasetto (1970) gave a rate of land subsidence which is
still more marked, 5 mm per year. The conclusion that the pronounced land subsidence in
Venice and the immediate vicinity is to a considerable extent artificial may be proved by
the fact that the average increase in mean sea level along the Mediterranean coasts is
generally markedly less, being, according to Polli (1962b), only about 1.5 mm per year.

The average height of Venice above the mean sea-level amounts to 50 cm. Urgent
measures are therefore required to save the city not only from final catastrophe in the
future, but also from disastrous floods which more and more frequently lay extensive
sections under water. Photograph 3 shows St. Mark's Square in Venice during a flood. As
a consequence of land subsidence, the frequency of floods caused by strong storm surges
and reinforced by astronomical tides and seiches is rapidly increasing. To compound the
problem the narrow strip of land which separates the lagoon from the open sea is be-
coming continually less capable of protecting the city from the effect of the largest storm
surges.

So far the sole explanation for the artificial subsidence in Venice has been sought in
the fact that numerous wells have been used in the city area, thus depriving the faults in
the ground of their natural pressure and causing a continuous and normally irreversible
compaction of the alluvial sediments of which the subsoil in the area is composed. Are
measures such as the prohibition of well-use and the regulation that all water must be
imported to Venice from the mainland sufficient to save the city? This is a question
which cannot be answered at the present time.

Frasetto says in his paper that 'the purpose of this presentation is to raise ... the
interest of the experts in the challenging scientific problem which represents a sort of
unique example of applied international research'. Can it really be possible, in the twentieth century with its marked increase in team work and international collaboration, that
this challenge cannot awaken the necessary response among oceanographers and other
concerned scientists of the world?

THE EUSTATIC FACTOR

The eustatic changes in mean sea level are generally ascribed to a large group of
phenomena such as folding of the seabed, sedimentation covering the sea floor and
melting or formation of continental ice masses. The first of the phenonema mentioned
above is more-or-less sporadic in character, while the second occurs continuously. However, an estimate of the effect of these factors upon changes in mean sea level is hardly
feasible at the present time. The phenomena as such have been mentioned by some
authors, but no efforts have been made so far to evaluate their contribution to the mean

Photograph 3. St. Mark's Square in Venice during a recent flood. The famous Basilica of St. Mark is threatened by water masses. (Photograph: Luigi Alberotanza, Venice.)

sea level of the world oceans. The picture changes completely when considering the investigation of the effect of fluctuations of the volume of continental ice masses upon the average height of the sea level in oceans and seas. These fluctuations deviate in character from the two other eustatically conditioned groups of phenomena. They are not as markedly sporadic as the folding processes, but neither are they continuous like sedimentation. They are also more easily measurable, since there are two different ways to study the phenomenon: (1) by purely cryological estimates; and (2) by examining the changes in mean sea level on a world-wide scale.

The cryological method is assuredly the more difficult. It must be kept in mind that estimates of the changes in the volume of the Antarctic ice sheet have a decisive effect upon the sea level, since these ice masses are by far the largest in the world. It has been estimated that if the total Antarctic ice sheet melted, the sea level in the oceans and seas would increase by approximately 100 m, causing a catastrophe of extreme dimensions.

Different estimates of the balance of the Antarctic ice masses have been made during the last few decades. According to King (1962) the majority of authors assume that there is an increase in the volume of the ice. Some suggest that the ice sheet is static, and only one author suggests that the ice volume is diminishing. On average, it must therefore be assumed that the ice masses in the Antarctic are in the process of growing. Since all researches on mean sea level show a distinct increase, the present situation in the Antarctic must be counterbalanced by the melting of the continental ice in the northern hemisphere, a process which is well established, being the consequence of the warming of the atmosphere. Whether and to what extent the warming of the oceans is a contributing factors to the increase in sea level is difficult to decide. This possibility must, however, be always kept in mind.

For how long a period of time the glacio-eustatic effect will continue in the future is a question which cannot be answered. The present situation may continue, the ice masses in the northern hemisphere melting and the sea level slowly increasing, but the possibility cannot be excluded that the present period is an interglacial one. If this is the case, the sea level will again begin to decrease, and the water will form into continental ice sheets over northwestern Europe and northern America.

The first author to give numerical values on the eustatic rise in mean sea level was Gutenberg (1941). Basing his computations on sea-level data for 69 stations in 22 different regions along the coasts, Gutenberg reached the conclusion that the increase amounts to 1.1 ± 0.8 mm per year. The probable mean deviation is fairly pronounced, which indicates that the result is not too accurate. At the time when the results of Gutenberg's computations appeared, the knowledge of the amount of water originating from the continental ice masses was very restricted. Only the results of Thorarinsson (1940) were available, according to which glacial fusion should cause a rise in mean sea level of 0.5 mm per year.

The efforts to solve this interesting and significant problem continued unceasingly. Kuenen (1950), combining different aspects, estimated that the eustatic increase in mean sea level is 1.2–1.4 mm per year. Dietrich (1954), in an extensive study of sea-level variations at Esbjerg, obtained the value 1.14 ± 0.28 mm per year, which, however, may be due to the vertical movement of the Earth's crust. Lisitzin (1958b) made an attempt to verify these results basing the computations on more extensive sea-level data. Instead of using a large number of stations referring to highly deviating periods, it seemed more appropriate to limit the study to a few stations, selected from those which were reliable and provided long series of data. The most interesting and rewarding stations in this respect were Brest in France and Swinemünde in the southern part of the Baltic coastal area. For Brest there were available monthly and yearly sea-level values for the years 1807–1835, 1846–1856 and 1861–1943, while for Swinemünde the series was practically complete from 1811–1943. Only seven months were missing in the years 1922 and 1923. In Table XXXVII are reproduced the average sea-level heights at these two stations for periods covering 10 and 20 years. Since the 20-year time-span deviates only slightly from the period of 18.61 years of the nodal tide the contributing effect of this astronomical tidal constituent is practically eliminated. The table shows very distinctly the increase in mean sea-level during the last decades, and in particular after the turn of the century.

The epoch for the start of the glacio-eustatic increase in mean sea-level is not well established. Some authors put forward as late a date as the 1930's. The turn of the century or possibly the last decade of the nineteenth century is, however, probably a better choice. The year 1891 has therefore been taken as the starting year of the increase. For the previous period of 80 years no pronounced deviations or tendency could be noted at Swinemünde. At Brest, however, as the values in Table XXXVII indicate, a weak increase in mean sea level was also apparent before the year 1891.

In order to determine the increase in mean sea-level since 1891 the following proce-

TABLE XXXVII

THE MEAN SEA LEVEL (CM) FOR PERIODS COVERING 10 AND 20 YEARS AT BREST AND SWINEMÜNDE

	Brest		Swinemünde	
	periods of 10 years	periods of 20 years	periods of 10 years	periods of 20 years
1811–20	446.4		−7.0	
1821–30	445.2	445.8	(−4.4)	−5.7
1831–40	−	−	−7.6	−6.0
1841–50	−	−	−7.5	−7.6
1851–60	−	−	−7.4	−7.4
1861–70	446.9	−	−6.5	−7.0
1871–80	449.3	448.1	−5.5	−6.0
1881–90	448.5	448.9	−6.7	−6.1
1891–1900	447.3	447.9	−4.4	−5.6
1901–10	447.8	447.6	−4.1	−4.2
1911–20	455.5	451.6	−2.0	−3.0
1921–30	453.4	454.4	0.4	−0.8
1931–40	455.2	454.3	−2.2	−0.9

dure has been chosen. With the help of the method of least-squares the increase in sea level at Brest and Swinemünde has been determined for the period before 1891 and for the period after. Starting from the assumption that for the period before 1891 the computed rate of sea-level increase is due to vertical movements of the Earth's crust, the rise in sea level caused by the continental deglaciation — and possibly also by other eustatic factors — since 1891 has been determined as the difference of the results computed for the two periods. The results obtained in this way were:

	For Brest	For Swinemünde
Before 1891	0.49 mm per year	−0.002 mm per year
After 1891	2.26 mm per year	0.70 mm per year

The increase caused by the fusion of the continental ice should thus be 1.77 mm per year for Brest and 0.70 mm per year for Swinemünde. The deviation between the rates of increase in mean sea level computed for the two stations is considerable. Therefore, other different contributing factors must be taken into account, since their effects cannot be completely excluded. There could possibly have been a change in the position of the zero height, especially during the former part of the last century when the sea-level heights were read from a tide pole. The contribution of the meteorological factors, which has not been eliminated from the yearly mean values, may be considerable. At Swinemünde in particular the effect of the wind upon the sea level may be pronounced, owing to the slight depth of the Baltic Sea. Nevertheless, for Brest also it can be established that by leaving out the year 1937, which was characterized by a very high yearly mean sea level, the rate of increase of the average sea level diminishes by not less than 5% i.e., from 0.226 cm to 0.214 cm per year.

In order to be able to verify the results obtained above and draw general conclusions based on more extensive data, a number of supplementary stations have been taken into consideration. Unfortunately, the greater number of stations with sea-level observations covering a relatively prolonged time-span before the year 1891 are situated in the Baltic. This sea basin is one of the least satisfactory for the study of the eustatic increase in sea level, not only because of the marked contribution of the wind effect, but also owing to the pronounced vertical movements of the Earth's crust in the northern parts of the sea area. The two stations Lyökki and Jungfrusund, which, in spite of the above-mentioned difficulties, represent the Baltic, are situated on the Finnish coast of the Gulf of Bothnia. They have both been in operation since 1858. In addition, sea-level data for Marseilles have been used. Since this station was installed in 1885, it has not been possible to determine the changes in mean sea level before 1891, and the assumption has been made that they are not of any importance. Finally, sea-level data have been utilized for Bombay where a station has been operating since 1878. For all these stations the rates of the changes in mean sea level were determined for the periods before and after the year 1891 on the basis of the least-squares method and, in order to obtain a comparison, also by the method developed by Gutenberg. This method implies that the variations in mean sea

TABLE XXXVIII

THE CHANGES IN MEAN SEA LEVEL (MM/YEAR) FOR DIFFERENT STATIONS BEFORE AND
AFTER THE YEAR 1891 DETERMINED ON THE BASIS OF THE LEAST-SQUARES METHOD (A)
AND THE METHOD OF GUTENBERG (B)

Station	Period	A	B	Period	A	B	Differences A	B
Lyökki	1858–1890	−6.8	−6.9	1891–1943	−6.0	−5.7	0.8	1.2
Jungfrusund	1858–1890	−4.8	−4.6	1891–1943	−3.6	−3.7	1.2	0.9
Swinemünde	1811–1890	0.0	−0.2	1891–1943	0.7	0.9	0.7	1.1
Brest	1807–1890	0.5	0.5	1891–1943	2.2	2.5	1.7	2.0
Marseilles	–	–	–	1891–1943	1.3	1.3	(1.3)	(1.3)
Bombay	1878–1890	0.5	0.7	1891–1943	1.5	1.6	1.0	0.9

level during a time-span covering n years may be determined with the help of the expression:

$$(a - b)/(n - m)$$

where a is the average height of the sea level during a period of m years at the beginning
of the time-span of n years, and b the corresponding height during a period of m years at
the end of the time-span. The probable deviation of this expression is at its minimum in
the case where $m = n/3$. The results are compiled in Table XXXVIII.

Table XXXVIII indicates that the deviations obtained by the two methods are not
very pronounced. On average, the eustatic increase in mean sea level has, according to
method (A) been determined at 1.12 ± 0.36 mm per year and the corresponding rate of
increase according to method (B) at 1.23 ± 0.41 mm per year. The departure between the
latter result and that given by Gutenberg himself is not pronounced. The probable deviation is still marked but less pronounced than that given by Gutenberg.

The eustatic increase in mean sea level in the oceans and seas brought about by the
continental deglaciation cannot, as mentioned above, be expected to be either strictly
regular or continuous. It is possible that periods characterized by the regression of the ice

TABLE XXXIX

THE AVERAGE RATE OF THE EUSTATIC INCREASE IN MEAN SEA LEVEL (MM PER YEAR)
ACCORDING TO DIFFERENT AUTHORS

Author	Computed results	Method
Thorarinsson (1940)	0.5 or more	cryological aspects
Gutenberg (1941)	1.1 ± 0.8	great amount of sea-level data
Kuenen (1950)	1.2 − 1.4	combining different aspects
Lisitzin (1958b)	1.12 ± 0.36	sea-level data for six stations
Wexler (1961)	1.18	cryological estimates
Fairbridge (1961)	1.12	sea-level data

masses have been interrupted by years during which the ice sheets have increased in volume. There are also some indications that there has been a slight retardation of the increase in mean sea level in recent years. Not even the average rate is therefore quite exact. Nevertheless, there is a pronounced trend in the data computed by different authors. Some of the more important results achieved by different methods are compiled in Table XXXIX.

The average value of the eustatic increase in mean sea-level determined by different authors thus amounts to 1.0–1.1 mm per year. In order to obtain the approximate rate of land uplift or land subsidence, this correction must be subtracted from the data representing the secular variation in sea level.

CHAPTER 7

SEICHES

Seiches are standing waves or stationary oscillations which occur in enclosed or semi-enclosed water basins such as lakes, gulfs, bays or harbours. The phenomenon was established for the first time by the Swiss physician F. A. Forel in the Lake of Geneva in 1869. The term seiches, which probably refers to the fact that owing to the oscillation a part of the shore becomes dry (sèche in French), was for a time used to characterize sea-level variations in lakes. Later it was also adopted to describe free stationary sea-level oscillations in other more-or-less enclosed water bodies. It was Forel's classic work (1895) on seiches in the Lake of Geneva which gave an impulse to the investigation of the phenomenon not only in other lakes but also in gulfs, bays and harbours.

Seiches in every enclosed or partially enclosed water basin have a natural period of free oscillation which depends upon the horizontal dimensions and the depth of the water body, and on the number of nodes of the standing wave.

If the length of the standing wave is marked in comparison with the depth of the water basin, seiches follow the theory of long waves in their behaviour. The velocity with which the wave progresses is thus given by the expression $\sqrt{gh}$, where g is the acceleration of gravity and h the depth of water. The wave arrives at the opposite end of the basin of lenght L after a time $L/\sqrt{gh}$. After reflection at this end the wave returns to the starting point in the same period of time. Thus the entire period is given by the expression:

$$T_1 = 2\,L/\sqrt{gh}$$

This formula was originally given by J. R. Merian in 1828. It represents theoretically the largest possible free oscillation with one nodal point. The natural period of oscillation for a completely enclosed rectangular water basin with constant depth is determined by the formula:

$$T_n = 2L/n\,\sqrt{gh}$$

where n is the number of the nodal points in the whole system. Antinodes occur at the opposite ends of the basin with the node or nodes in the central areas.

A schematic picture of a bi-nodal seiche oscillation is given in Fig. 36.

In the cases where the height of the standing wave, corresponding to the vertical displacement η of the sea level, is not very slight in comparison with the depth of the water, this height also exerts an effect on the velocity of the wave. The period of the oscillation decreases and may be computed on the basis of the expression:

$$T_s = T_1 \left[1 - 3/2 \left(\eta/h\right)\right]$$

This formula may be used exclusively in connection with enclosed basins. For gulfs and bays which have an opening at one end, a nodal point always occurs at this end, since the water masses outside the gulf or bay also participate in the horizontal oscillation. The result is, therefore, that the period of uni-nodal free oscillation in a semi-enclosed sea basin is twice as long as the period in a completely enclosed basin with similar horizontal and vertical dimensions:

$$T_1 = 4L/\sqrt{gh}$$

This formula is, however, applicable only in the cases where the opening at one end of the basin is narrow. If the opening is large compared with the length of the basin, a correction, the so-called mouth correction, must be applied to the period formula. It has been estimated that the period of oscillation increases by 18%, when the breadth of the mouth is 1/5 of the length of the sea basin and that this increase amounts of 37%, when the width of the opening is equal to the length of the basin.

Forel was able to establish that the observed period of oscillation in the Lake of Geneva was in rather poor agreement with Merian's formula. Different methods were developed in order to determine the actual periods of free oscillations in the particular basins. All these methods were based on the hydrodynamic equations of motion, but they differed considerably in the procedures for finding the solution to the problem. The most frequently applied method is probably that given by Defant (1918). According to this method a solution may be reached for the most complicated configurations of the basin by a stepwise integration, using the finite difference method. Besides the period of oscillation, Defant's method offers the possibility of determining the vertical departures of sea level from the average height and the horizontal motion of the water particles. It is notably in this respect that the method of Defant is preferable to the other methods, in spite of the fact that it requires a considerable amount of work. Among the other methods mention may be made of those developed by Chrystal (1905), Proudman (1914), Ertel (1933) Hidaka (1936) and Neumann (1941). For a more detailed description of these methods reference may be made to e.g., Defant (1961, pp. 161–173).

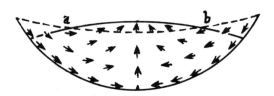

Fig. 36. A schematic picture of a bi-nodal seiche oscillation.

Hidaka has also in a number of more recent publications paid particular attention to the problem of seiches and its theoretical solutions.

In natural conditions seiches are generated by abrupt changes in the direction or velocity of the wind, by gradients in atmospheric pressure or by co-oscillation with an adjoining water body. Even relatively weak external forces may be able to generate a prolonged series of gradually damped oscillations, which in some cases may reach considerable proportions. This occurs in particular in situations where the period of the generating forces is close to the natural oscillation period of the basin.

Sea-level records have proved that all semi-enclosed water basins execute oscillations with a nodal line at the opening of the gulfs and bays. This type of 'disturbance' in the smooth course of the recorded curves occurs quite frequently. Seiche-like oscillations were already ascertained at the close of the last century and during the first decades of this century in a great number of different basins around the European coasts. In this connection reference must be made to a number of bays along the German North Sea coast and to waters in the transition area around Denmark. In the Mediterranean Sea the number of localities characterized by the occurrence of seiches is highly pronounced. Von Sterneck (1914) was able to establish, in an extensive publication on the seiches in the Adriatic Sea, that there exists an adequate correspondence between the observed period and that determined on the basis of Merian's formula. The free oscillation in the bays of the Adriatic Sea generally have a period not exceeding half an hour. Only in Trieste are seiches with an average period of 3.2 hours sometimes observed. This type of oscillation always causes considerable variations in sea level, the average range for 15 cases being approximately 75 cm. These seiche oscillations cover the whole Gulf of Trieste and are limited from the open sea by a line extending from Capo Salvadore to a point situated somewhat to the west of Grado. This line corresponds to the nodal line of the oscillation system. The application of Merian's formula, taking into account a correction for the opening corresponding to 1.29, results in a period of 3.1 hours, being thus in fairly good agreement with the observed period (Caloi, 1938). In addition, a part of the Adriatic Sea situated to the west of the nodal line participates in the oscillation, since the tidal gauge at Falconera, having a position approximately 30 km to the west of Grado, shows a distinct antinode with phase opposite to that of Trieste. Shorter oscillations with a period of 0.78 hours or less, recorded in Trieste, are seiches occurring in the Bay of Muggia and in the harbour.

The period of seiches in the whole basin of the Adriatic Sea was investigated by Polli (1961). The results were based on actual observations of the sea-level variations and on computations which had been previously made, taking into account the correction required by the opening. The results, showing the best agreement for uninodal seiches, are given in the following:

	Observed periods	Computed periods
Uni-nodal seiches	21 h 20 min	21 h 50 min
Bi-nodal seiches	12 h 11 min	10 h 25 min
Tri-nodal seiches	7 h 34 min	7 h 00 min

Of considerable interest are the standing oscillations which were observed in the 'haffs'. These are extended basins of slight depth connected with the open sea by one or several openings. In the Baltic Sea area typical examples of this kind of basin are the Frisches Haff and the Kurisches Haff. In the Frisches Haff — with a length of 90 km and average depth less than 3 m — standing oscillations occur with periods of 8.0 hours and, less frequently, of 5.0 hours. According to Defant (1961) the former period probably refers to the oscillation of the entire haff, while the latter period may be that of bi-nodal seiche. The Kurisches Haff — length 85 km and average depth 4 m — shows, as a rule, an oscillation period of 9.2 hours, probably corresponding to that of the uni-nodal seiche and, in addition, a period of 4.1 hours, which according to Defant must be interpreted as that of the bi-nodal seiche.

In common with a considerable number of different problems concerned with sea-level changes, seiches have also been studied in the Baltic Sea in great detail. Witting (1911) was probably the very first to determine an approximate value for the oscillation period of the Baltic proper and the Gulf of Finland which forms an immediate continuation of the former basin. The basis for the computations were a number of theoretical assumptions and different simplifications of the bottom configuration of the basins, resulting for the separate cases in values which deviated considerably from each other. Nevertheless, the final conclusion drawn by Witting was that the period of the uni-nodal oscillation varied between 28 and 31 hours, a result which must be considered to be a fairly good approximation. Thirty years after Witting, Neumann (1941) again paid attention to the problem. Neumann made an attempt to solve the problem in two different ways: on the one hand on the basis of sea-level records for a number of selected cases characterized by a pronounced oscillation, and on the other hand with the help of extensive computations involving the numerical methods derived by Defant and Hidaka and already referred to here. This part of Neumann's study, which was by far the most significant, required an exact knowledge of the horizontal dimensions and the depth of the sea basins. The correspondence between the individual results achieved by Neumann using the two different procedures was surprisingly good. Thus the average uni-nodal oscillation period for the system Baltic proper—Gulf of Finland determined on the basis of sea-level records amounted to 27.6 hours, while the theoretical calculations resulted in 27.5 hours. The principal condition for the reliability of this value is, however, the assumption that the Baltic Sea is enclosed in the southwest at the Fehmarnbelt, which of course does not correspond to the actual situation.

The results obtained by Neumann are highly satisfactory. Nevertheless, attention must be paid to a few factors influencing the data. Only six readings a day were used from the sea-level records for the Swedish, Finnish, Estonian and Latvian stations. Since the amplitudes of the oscillation are at their highest and the phenomenon as a whole thus best developed in the Gulf of Finland, the determination of the period was based on what are comparatively incomplete data. Moreover, the computation of the period was based on a time-span between the first and the last maximum investigated in each case, also paying attention to the number of intermediate extremes. This procedure may be the cause of considerable deviations, since local irregularities in sea-level variations are frequent, in particular towards the end of the oscillation period when the range of variations has already decreased. In fact the oscillation period given by Neumann for the different cases varies between 26 and 28 hours.

The theoretical calculations of the oscillation period performed by Neumann are very instructive. The complete agreement between the results achieved in this way is, without doubt, an excellent proof of the reliability of the two methods, i.e., those of Defant and Hidaka. However, Neumann himself has pointed out that the irregular configuration of the Baltic basin and especially the difficulty in determining exactly the confines of the oscillation area involve a certain instability in the solution of the problem. It has already been mentioned that Neumann assumed, in his first attempt to compute the oscillation period, that the Baltic Sea is closed at the Fehmarnbelt. Including the shallow Bight of Kiel in the oscillation system, Neumann arrived at an oscillation period of 29.5 hours on the basis of the method of Defant, while the method of Hidaka resulted in a period which was 0.2 hours shorter. The pronounced agreement with the results obtained from sea-level records is no longer valid. This example shows very distincly the dependence of the oscillation period on the dimension of the sea basin and gives a good picture of the difficulties involved. It may be added that Neumann's tables show numerous deviations between the computed amplitudes and the corresponding recorded values. The position of the nodal line is also not quite the same for the theoretical results and for the actual cases.

These deviations proved that the problem was by no means solved, if accurate results were the main aim of the research work. It had to be approached once more, basing the determination of the period on a great number of sea-level records (Lisitzin, 1959d). The difficulty of obtaining sea-level data from the countries situated in the southern and eastern parts of the Baltic (these data have, as a rule, not been published since the beginning of the Second World War) meant that the research had to be planned and carried out with sea-level records from the Finnish stations constituting the principal and essential part of the work. Only in a restricted number of cases could the data for the Finnish coasts be compared with a few representative results for the southern parts of the Baltic. From the German coast, some data from Strande, situated outside Kiel were available and from Poland data from Wladislawowo, located off the peninsula of Hel, from Kolobrzeg and from Swinoujscie. The position of the stations is given in Fig. 37.

As a starting point for the research the hourly sea-level records for Hamina, which since the end of World War II has been the easternmost Finnish sea-level station in the Gulf of Finland, were chosen. From the records for the years 1952–1954 a number of cases were selected which showed a distinct and relatively regular oscillation with a period of roughly 26–28 hours. Generally these cases were fairly frequent, but only the results for series which fulfilled the following three conditions were taken into consideration:

(1) The number of consecutive, well-developed oscillation cycles had to be at least four.

(2) A rough estimate of the range of the oscillation had to result in at least 12 cm.

(3) In order to ascertain that the oscillation represented a uni-modal seiche in the

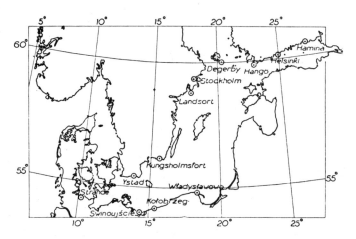

Fig. 37. The position of the sea-level stations used for the study of seiches.

system represented by the Baltic proper—Gulf of Finland, the records for the Swedish sea-level station Ystad, situated in the southern part of the Baltic Sea, were investigated to check that a more-or-less distinct inverse oscillation could be proved to have occurred there.

In addition, the possibility was taken into account that a new seiche oscillation could have been induced, influencing the computed length of the period. It was therefore checked, in all the cases with signs of an increase in range during the selected period, that no displacement of the extreme values had taken place.

It was established that uni-nodal seiches corresponding to the above-listed stipulations were by no means rare in the oscillation system Baltic proper—Gulf of Finland. Not less than 9% of all sea-level records for the 3 years examined, resulting in a cumulative time of more than 3 months, showed a pronounced uni-nodal oscillation. The number of cases with a more restricted average amplitude and with fewer periods was considerably higher. The distribution of the cases over the years was, however, not very even. For the year 1952 there could be utilized no less than 9 cases corresponding altogether in 51 periods, for the year 1953 only 3 cases with a total number of 12 periods and for the year 1954 5 cases covering 25 oscillation periods. These cases correspond in round figures to 15, 4 and 7% respectively for these 3 years.

As soon as the cases were selected the harmonic constants were computed. Starting from different estimated oscillation periods, those resulting in maximum amplitudes were determined with great accuracy. The results of these computations are collected in Table XL. This table shows that although the periods examined cover a time from 25 to 27 hours, the period connected with the extreme values of the amplitude generally varies between 26.0 and 26.4 hours. In fact, there are in Table XL not more than two exceptions from the above results. The computed period for the oscillation recorded in January

TABLE XL

THE HARMONICALLY DETERMINED AVERAGE AMPLITUDES (CM) OF THE UNI-NODAL SEICHE OSCILLATION AT HAMINA FOR DIFFERENT LENGTHS OF OSCILLATION PERIOD

Length of the period (hours)	1952								
	Jan.16–29 (12 periods)	Feb.10–14 (4 periods)	Feb.16–21 (5 periods)	March 16–21 (5 periods)	March 26–31 (5 periods)	April 2–8 (5 periods)	April 15–20 (5 periods)	July 31–Aug. 5 (5 periods)	Nov.18–23 (5 periods)
25.0	7.9	17.5	–	8.9	–	–	–	–	–
25.5	8.8 *	18.1	–	9.1	7.0	–	–	–	8.3
26.0	7.4	18.7	7.3	9.3	7.9	8.9	8.9	5.6	9.0
26.2	6.6	18.7	7.5	9.1	7.6	8.8	9.4	5.8	8.5
26.4	4.7	18.3	6.5	9.0	7.7	9.2	9.1	5.5	7.9
26.6	–	–	–	–	–	9.3	–	–	–
27.0	–	–	–	–	–	9.2	–	–	–

TABLE XL (continued)

Length of the period (hours)	1953			1954				
	March 23–27 (4 periods)	Oct.22–26 (4 periods)	Dec. 2–6 (4 periods)	Jan.15–20 (5 periods)	June 17–July 1 (4 periods)	Oct.14–21 (4 periods)	Nov.10–15 (5 periods)	Nov.27–Dec.1 (4 periods)
25.0	–	–	–	–	6.4	–	–	11.7
25.5	–	7.6	–	–	7.0	18.5	26.4	12.5
26.0	12.1	7.8	26.9	14.9	7.1	18.8	28.2	12.6
26.2	13.1	7.6	28.2	15.2	6.8	18.9	27.8	11.9
26.4	13.3	7.2	26.8	15.0	6.7	18.7	28.3	11.6
26.6	12.8	–	–	–	–	17.1	26.8	–
27.0	–	–	–	–	–	–	–	–

* Italicized values denote maximum amplitudes.

1952 was distinctly shorter than average, i.e., approximately 25.5 hours. The number of
consecutive periods utilized for the determination of the maximum amplitude was, however, in this case exceptionally high, being 12 against the average of 4 or 5 periods. It is
therefore by no means excluded that a certain, even if weakly pronounced, displacement
of the seiche oscillation occurred during this relatively prolonged time-span, influencing
the results accordingly. The other exception from the general rule in Table XL is the
oscillation recorded at the beginning of April 1952. The maximum amplitude corresponds
in this case to a period of 26.6 hours. The departure from the more normal period is thus
rather insignificant. Moreover, the table shows that the maximum of the amplitude is, as a
rule, not accentuated, indicating that occasional disturbances may influence the final
results. Nevertheless, it seemed adequate to consider 26.2 ± 0.2 hours as the average
length of the oscillation period derived on the basis of 17 recorded cases in the system
Baltic proper — Gulf of Finland.

In order to check the results obtained above, the sea-level records for Władisławowo
were considered for four cases. The computed data showed that the oscillation period
giving the maximum amplitude was in three cases noted between the limits of 25.7 and
26.4 hours. In addition, the differences between the phase angles for the two stations
indicated in these three cases a pronounced inversity of the oscillation, the phase angle

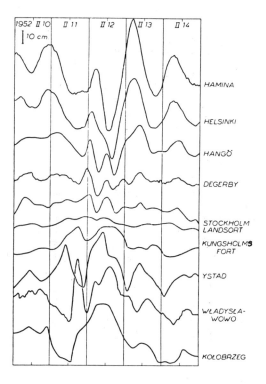

Fig. 38. Sea-level variations, February 10–14, 1952 (Lisitzin, 1959d).

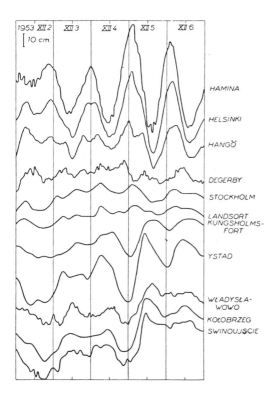

Fig. 39. Sea-level variations, December 2–6, 1953 (Lisitzin, 1959d).

being close to 180°. The amplitudes for Władisławowo were weak, only 22–35% of those for Hamina, which implied that the results for the former station were less reliable. In the fourth case examined, the coincidence was not wholly satisfactory. On the one hand, the oscillation period for Władisławowo was markedly shorter than that for Hamina, amounting to only 23.8 hours, while on the other hand the stipulation for the inversity of the oscillation at the two stations was only approximately fulfilled. It must therefore be assumed that a factor of more-or-less local character had disturbed the regular course of the phenomenon and influenced the results.

Fig. 38 and 39 give a graphical presentation of the data examined above. These figures refer to the time-span of February 10–14, 1952 and December 2–6, 1953, respectively. In addition to the sea-level data for Hamina and Władisławowo the figures give the hourly records for three more Finnish sea-level stations: Helsinki, Hangö (Hanko) and Degerby. The curves representing the Swedish stations Stockholm, Landsort, Kungsholmsfort and Ystad and the Polish station Kołobrzeg are based on 6 readings daily and the curve for Swinoujscie in Poland (Fig. 39) on 12 readings daily.

The figures show that the oscillation is highly pronounced for Hamina and Helsinki. Progressing to Hangö it may easily be noted that the phenomenon as a whole has a similar character, but the amplitudes show a considerable decrease and there occur some supplementary maxima. For Degerby and Stockholm the general features of the oscillation are no longer very distinct, and for Landsort the phenomenon is hardly perceptible owing to the proximity of this sea-level station to the position of the nodal line. The inversity of the phenomenon in the southern part of the Baltic Sea is indisputable, but the curves in Fig. 38 and 39 prove that the number of different marked local and regional perturbances is quite pronounced, rendering the exact determination of the oscillation period not only difficult but practically impossible.

Krauss and Magaard (1962) made a renewed attempt to determine theoretically the oscillation period of the system Lübeck–Gulf of Finland. These authors computed not only the period of the uni-nodal seiche oscillation, but also those periods corresponding to seiches characterized by more than one nodal line. The results, which are based principally on the same method as that followed by Neumann, are the following (T_n referring to an oscillation in the concerned system with n nodal lines):

T_1 = 27.4 hours T_5 = 8.1 hours
T_2 = 19.1 T_6 = 6.9
T_3 = 13.0 T_7 = > 6
T_4 = 9.6   5 hours < T_8 < 6 hours.

For the uni-nodal oscillation the period is thus practically the same as that computed by Neumann, which of course is by no means surprising, since the method of computation was on the whole the same. The discrepancy between the result achieved by Krauss and Magaard and the oscillation period determined on the basis of the recorded data is not so easy to explain. However, it must always be kept in mind that the exact effect of friction upon the period is so far not known. In this connection it may be appropriate to mention that Laska (1969), on the basis of the analysis of periodic sea-level variations according to the power spectrum method, determined that the uni-nodal oscillation period in the system Baltic proper–Gulf of Finland amounts to 26.6 hours. These results are based on the records of two sea-level stations, Sopot and Kołobrzeg, situated on the Polish coast.

The above-given periods of higher oscillations − T<10 hours − have not been computed before. Krauss and Magaard (1961) showed that they occur frequently in the recorded data.

An attempt was also made to determine the oscillation period in the system Baltic proper–Gulf of Finland, including the Bight of Kiel, by Lisitzin (1959d). Since the amount of data for Strande was restricted, covering only one year, it was not possible to achieve final results. One case extending from January 7–12, 1955 may, however, be mentioned. The harmonic analysis resulted in this case in a period of 28.8 hours for Hamina and 27.7 hours for Strande. The deviation between the two periods is too high to allow a general conclusion. Nevertheless, it is symptomatic that the period in this case also is shorter than that computed by Neumann, which amounted to 29.3–29.5 hours.

Fig. 38 and 39 show that on following the Finnish coast from east to west a slight retardation is visible in the occurrence of the extreme sea-level heights. This retardation can only be explained on the basis of a transverse oscillation in the Gulf of Finland. Since Harris' (1904) investigations of this phenomenon, it has been known that in basins with considerable width standing waves may occur not only along the longitudinal axis but also along the transversal axis. Since interference arises between the two waves, they form a rotational wave which, according to Harris, may also be called an amphidromic wave. In order to obtain a concept of the phase lag occurring in the Gulf of Finland, the harmonic constants were computed for the oscillation period of 26.2 hours at Hangö for all the cases given in Table XL, with the exception, however, of the case referring to April 1954, for which records at Hangö are missing. In Table XLI these constants and the corresponding data for Hamina are reproduced. The ratios of the amplitudes and the departures between the phase angles are also to be found in the table. The general impression obtained from the table is the lack of conformity. Thus the ratios between the amplitudes for Hamina and Hangö vary from 1.7 to 10.1. The latter ratio is, however, quite exceptional. According to the theoretical computations made by Neumann and by Krauss and Magaard this ratio should amount to approximately 2.5. The limits for the fluctuations of the difference in phase are $+33°$ and $-25°$. All the cases where the extreme sea level appears at Hangö earlier than at Hamina must be left out of consideration in this connection, since they cannot be the consequence of a transverse oscillation, but must be due to disturbances of more local character. The remaining cases are also probably influenced by perturbating effects. The average value for the positive phase difference in Table XLI is $12°$, which corresponds roughly to a time lag of 55 min. This value is only approximate, since the number of cases does not allow a more exact determination. The value is about 35% too low, if compared with the time lag determined by Neumann. This author has computed that the retardation between Koivisto, situated in the Gulf of Finland about 100 km to the east of Hamina, and Hangö amounts to 1.8 hours. Starting from this value we obtain by interpolation a retardation of 85 min between Hamina and Hangö.

The general outlines of the present knowledge of seiches in the Baltic are incomplete, if no attention is paid to the system Baltic proper–Gulf of Bothnia. It is true that seiches are observed very seldom in this oscillation system. The oscillation period computed by Neumann amounted to 39.1 hours. Krauss and Magaard (1962) also renewed the computations for this oscillation system, obtaining for the oscillating water body extending from Lübeck to the innermost parts of the Gulf of Bothnia the following results:

$T_1 = 39.4$ hours $T_5 = 9.4$ hours
$T_2 = 22.5$ $T_6 = 7.3$
$T_3 = 17.9$ $T_7 = 6.9$
$T_4 = 12.9$

The general impression of the above-given oscillation periods compared with those for the oscillation system Lübeck–Gulf of Finland is that they decrease more rapidly with

TABLE XLI

HARMONIC CONSTANTS FOR HAMINA AND HANGÖ CORRESPONDING TO AN OSCILLATION
PERIOD OF 26.2 HOURS

	Hamina		Hangö		Ratios of the ampli- tudes	Phase angle differ- ences
	ampli- tude (cm)	phase angle	ampli- tude (cm)	phase angle		
1952						
Jan. 16 − 29	6.6	86°	3.9	108°	1.7	22°
Feb. 10 − 14	18.7	322°	4.7	355°	4.0	33°
Feb. 16 − 21	7.5	233°	3.7	235°	2.0	2°
March 16 − 21	9.1	210°	1.4	202°	6.5	−8°
March 26 − 31	7.6	159°	3.3	146°	2.3	−13°
April 2 − 8	8.8	108°	2.9	117°	3.0	9°
April 15 − 20	9.4	258°	3.9	269°	2.4	11°
July 31 − Aug. 5	5.8	64°	2.8	78°	2.1	14°
Sept. 18 − 23	8.5	332°	3.3	331°	2.5	−1°
1953						
March 23 − 27	13.1	288°	5.9	274°	2.2	−14°
Oct. 22 − 26	7.6	130°	3.1	137°	2.5	7°
Dec. 2 − 6	28.2	100°	2.8	103°	10.1	3°
1954						
Jan. 15 − 20	15.2	235°	3.8	233°	4.0	−2°
Oct. 14 − 21	18.9	250°	5.1	264°	3.7	14°
Nov. 10 − 15	27.8	176°	7.0	151°	4.0	−25°
Nov. 27 − Dec. 1	11.9	27°	5.4	32°	2.2	5°

increasing number of nodal lines than in the latter case.

This comprehensive presentation of seiches in the Baltic has been given mainly in order
to show the deviations existing between the theoretical results and those based on sea-
level records. Similar investigations have assuredly been made also for other basins, but
one sea region discussed in more detail may be sufficient to illustrate the general features
of the problem.

CHAPTER 8

TSUNAMIS – EARTHQUAKES AND MEAN SEA LEVEL

TSUNAMIS

A tsunami is a long surface wave which is caused by a submarine earthquake, by volcanic eruption on the sea bottom or, in recent times, by powerful atomic bombs exploded over the oceans. Tsunami is a Japanese word: 'tsu' means harbour and 'nami' wave. This name is thus highly appropriate, as tsunamis are principally coastal phenomena, bringing considerable damage to harbours, particularly those situated in shallow bays with wide openings towards the sea. Although the term tsunami is now internationally accepted, the English term 'seismic sea wave' may also be used. The term 'tidal wave' which is sometimes encountered in the literature is not correct, since a tidal wave refers exclusively to the periodically occurring astronomical tides. The use of this term may therefore be highly confusing and should be avoided. A large wave of the character of a storm surge induced by a violent meteorological effect should also not be denoted as a tsunami.

One of the most tragically famous earthquakes, followed by a disastrous wave, was the earthquake which practically destroyed Lisbon on November 1, 1755. The epicentre of the earthquake was probably situated approximately 100 km to the west of Lisbon. The height of the succeeding wave is estimated to have been 13 m. This wave completed the destructive work of the earthquake; the death toll in Lisbon was reported to have been 50,000–80,000 persons.

The frequently described volcanic eruption of Krakatoa which occurred on August 27, 1883, produced a tsunami wave of at least 15 m in height. Some authors give as high values as 35 m, which is probably highly exaggerated. Some 36,000 inhabitants along the coasts of the islands in the Sunda Strait area between Java and Sumatra lost their lives during the inundations caused by the tsunami wave.

From the list of tsunamis which have occurred during the twentieth century and which therefore are more properly documentated may be mentioned the wave which arose as the consequence of the Chilean earthquake of May 23, 1960. This wave originated off the coast of southern Chile and crossed the whole Pacific Ocean, reaching the coasts of Japan. The wave height along the South American coast in the neighbourhood of the origin was more than 10 m. In Chile not less than 900 persons were killed, in the Hawaiian region about 60 and in Japan more than 100 persons.

The destructive power of tsunamis is enormous. It has, for instance, been reported that mature trees have been completely wrecked by tsunami waves and wooded beaches have presented after the disaster an almost impenetrable tangle of uprooted, twisted, broken and shattered trunks.

The considerable damage and loss of life from tsunamis has forced scientists in the affected regions to intensify their research work. One major step was the establishment of an effective tsunami warning system (pp. 244–245). The countries most strongly affected by tsunamis and therefore most interested in the problem are Japan, the United States of America and the Soviet Union. All these countries command their own warning centres which, particularly since the Chilean earthquake and tsunami of 1960, have strengthened their international collaboration. Each centre has at its disposal a number of stations which report seismological data whenever earthquakes are observed, and a great number of stations reporting wave amplitudes by means of sea-level changes as soon as the amplitude of these variations (peak to trough) exceeds one metre. The purely theoretical investigations on tsunamis have also been markedly intensified during the last decades, particularly in the three countries which may be expected to suffer the greatest damage.

The description of the characteristics and mechanism of the earthquakes lies beyond the scope of this presentation. On the contrary, the relationship between different types of earthquakes and tsunamis is significant for theoretical and practical research. Is is self-evident that the zones with a pronounced earthquake frequency are also the regions with a marked occurrence of tsunamis. The most prominent area in this respect is the Circum-Pacific region which is considered to account for almost 80% of all recorded tsunami cases. The records for the far bygone days are, of course, not very complete. For the first half of the twentieth century it has been determined that the average frequency of tsunamis amounted to 1.4 cases per year (Heck, 1947).

Not all submarine earthquakes produce noteworthy tsunamis. It seems that vertical motion of the sea floor resulting in a local elevation or depression is a prerequisite. In this way arise condensational waves, usually denoted as seaquakes. This phenomenon may be considered as the first phase of tsunamis. The second phase is characterized by the propagation of a train of gravitational oscillation waves which under certain conditions proceed in all directions from the submarine epicentre with a velocity depending on the depth of the ocean. Since the wave length of a tsunami may be highly pronounced, it frequently passes unnoticed by observers on board vessels at sea. As soon as a tsunami wave approaches a large island or the continental coast, the added effect of refraction, interference and resonance may produce a considerable local increase in wave height. Tsunami waves may, as a third phase of their development, cause coastal bays to develop stationary sea-level oscillations, the period of which is determined by the shape and dimensions of the bay. In this way arise marine seiches which in some cases may last for several days.

A significant, although partial, problem in tsunami investigations is the magnitude classification. Iida (1963) studied the relationship between the tsunamis around Japan and the magnitude of the corresponding earthquakes on a statistical basis. Resulting from this research, the following empirical formula was achieved, on which the Japanese tsunami warning system is based:

$$m = (2.61 \pm 0.22) M - (18.44 \pm 0.52)$$

In this equation m denotes the magnitude of the tsunami and M that of the earthquake according to the Gutenberg–Richter scale. The earthquake shock must be of a magnitude greater than 7.3, in order to cause a tsunami characterized by noteworthy shore damage ($m>0$). For earthquake shocks greater than 7.8 a considerably more disastrous tsunami wave causing coastal damage and loss of human life may be expected ($m>2$). Very severe destruction ($m>3$) occurs in connection with earthquakes of the magnitude of 8.2.

The magnitude of the tsunamis, their heights and devastating consequences are characterized by the features listed in Table XLII.

TABLE XLII

THE MAGNITUDES AND CHARACTERISTICS OF TSUNAMIS

Magnitude	Height of tsunami waves	Damage potential
−1	50 − 70 cm	no damage
0	1 − 1.5 cm	very little damage
1	2 − 3 m	shore damage
2	4 − 6 m	some inland damage and loss of life
3	8 − 12 m	severe destruction over 400 km of coast
4	16 − 24 m	severe destruction over 500 km of coast

Besides depending on the magnitude of the shock, the magnitude of the tsunami is dependent on the focal depth of the earthquake and decreases with increasing depth. According to Iida this relationship may be expressed by the formulae:

$$M_o = 6.5 + 0.008 H$$

and

$$M_d = 7.75 + 0.008 H$$

In these formulae H is the focal depth (in km) of the earthquake, M_o the limiting magnitude beyond which tsunamis are invariably generated and M_d the corresponding limit for more disastrous tsunamis.

The investigations of Matuzawa (1936) showed that the epicentre of the principal shock is generally situated at a corner of the after-shock area. The shape of this area is frequently elliptical. The ratio between the major axis a and the minor axis b of the ellipse increases with the magnitude of the earthquake. Hatori (1963) gave the following formula which was derived on a statistical basis:

$$b/a = AM + B$$

where $A = -(0.13\sim0.28)$ and $B = 1.69\sim2.56$. In addition, Hatori was able to show that the wave heights emitted in the directions of the two axes of the elliptical after-shock area are inversely proportional to their lengths, while the periods of the waves are directly proportional to the lengths of the axes. A significant factor in this respect is, however, the bottom topography of the region surrounding the shock area and the pattern of the deformation of the sea floor as a consequence of the earthquake. As a rule, adequate data are completely lacking in this respect. Matuzawa (1937) gave an example where the initial tsunami wave rose along the coast to the east, but receded to the west. The author tried to explain this phenomenon on the basis of an assumption that there must have been an upward motion to the east, but a downward motion to the west at the sea bottom within the region of seismic activity.

It is not yet within the limits of possibility to give detailed descriptions of the characteristics of the shape of a tsunami wave arriving at the coast. Two comprehensive groups of factors must always be taken into consideration in this connection. On the one hand, attention must be paid to the direction in which the tsunami wave approaches the coast and the energy associated with it. On the other hand, the topography of the sea bottom, the extent, depth and character of the shelf and the configuration of the shore-line are of considerable importance.

From the point of view of sea-level investigations the study of tide-gauge records before and during the arrival of the tsunami wave is a unique way of obtaining a concept of the phenomenon. These records, the tide gauges generally being installed in the coastal areas, give rather complicated pictures which frequently have very little similarity with the phenomenon as seen in the off-shore regions. As an example it may be mentioned that seiche oscillations frequently interfere with the successive trains of tsunami waves which themselves may arrive by different paths. Theoretically the head of the tsunami wave will, however, for the long periods involved in the process, follow the law of wave motion in shallow water. This implies that the time of propagation may be determined with the help of the velocity equation:

$$v = \sqrt{gh}$$

where h is the depth of the sea and g the acceleration of gravity. For the sea-level oscillations occurring later in the tidal records the relationship between velocity and period must be taken into account. In addition it must always be kept in mind that the recorded features of the tsunami wave will always be superposed by the tides and also by a possible meteorological contribution to the sea level.

The energy inherent to the tsunami wave may be approximately estimated from sea-level records. The formula to be used in this connection is:

$$E = \pi \rho g v R \Sigma \alpha^2 T$$

where ρ is the density of the sea water, R the distance from the zone of origin of the tsunami, α the amplitude of the oscillation and T the half period. As above, v is the velocity of the motion. The energy is supposed to be emitted symmetrically in all directions. Numerous computations and estimations of the energy of tsunami waves determined by different authors have shown that the largest and most powerful tsunamis reach an energy of 10^{24} erg and the weakest 10^{20} erg. The energies of the tsunamis are thus of the order of one-tenth to one-hundredth of the energies of the earthquakes. The energy of the tsunamis are, moreover, a function of the type of sea-floor dislocation caused by the earthquakes.

Frequently the opinion has been expressed that a tsunami at the coast is heralded by a receding sea level. There are stories of victims who during the recession of the sea water have gone far out from the shore to pick up shells and have been trapped by the arrival of the tsunami wave. Such cases have certainly been reported; nevertheless, mareographic records connected with tsunamis have shown that the first obvious disturbance is just as likely to be an increase in sea level. This indicates the marked complexity of the tsunami phenomenon as a whole.

In spite of the difficulties in giving a correct description of the development and progression of a normal tsunami wave, some general features may be observed fairly frequently. Soon after the occurrence of the shock progressive long waves are spread annularly as unbroken ridges in all directions from the epicentre. In the immediate vicinity of the epicentre fairly large amplitudes of the tsunami waves have been observed. If the distance from the shock area to land is considerable, the wave height may gradually decrease, but as soon as shallow water is entered the front of the tsunami wave becomes steeper and then impending. In the cases where the shore is low-lying ground, the wave sweeps over it with an enormous force. The height of these waves is not always a function of the magnitude of the earthquake. As has already been mentioned, it also depends on the inclination of the sea floor in the neighbourhood of the coast and the configuration of the shore-line. The highest amplitudes are thus reached in the inner parts of gradually narrowing bays. Personal observations and reports about the height of tsunami waves are by no means always reliable, and are frequently exaggerated. The estimates may vary for the same case within limits as great as 5—25 m. It may, moreover, be significant to know whether the crest height of the wave has been referred to the mean sea level prevailing before the arrival of the disturbance, or to the succeeding trough; i.e., if the height difference has been given as amplitude or as range of the wave. In any case there is no doubt that the heights may be considerable, and the examples mentioned at the beginning of this chapter are by no means exceptional.

The first tsunami wave is generally followed by several secondary waves, which are frequently of less pronounced height. The number of these secondary waves may vary within considerable limits, from a few to far more than ten. It has, for instance, been reported that the earthquake of Lisbon in 1755 was followed by 18 tsunami waves.

Earthquakes of sufficient magnitude on the two sides of the Pacific Ocean generate tsunami waves which frequently are also observed on the opposite coasts. This fact distinctly shows that the waves may travel distances exceeding 10,000 km. Taking into consideration the time interval between the shock and the arrival of the tsunami at the coast, average velocity values of the magnitude of approximately 700—800 km per hour can be determined.

Taking into account the velocity of tsunami waves, it is possible to draw a chart representing the propagation of the disturbance over the ocean from the place of its origin. This pattern is called a refraction diagram and it is of great assistance in obtaining information on the travel time. On the other hand, it is possible also, to construct a chart on which a tsunami wave is propagated from the points of observation on the coast towards the deep sea with the purpose of detecting the epicentre of the earthquake. Such a chart is called an inverse refraction chart.

Finally, a few words may be dedicated to a man-made gravity wave, which arose as a consequence of the nuclear explosion at Bikini, Marshall Islands, in 1956 (Van Dorn, 1961). The dispersion of the disturbance was of the same character as for waves associated with large tsunamis and in agreement with the theory that the dispersion of the wave system is not dependent on the character of the origin. Regarding the rate of decay of the amplitude, it was necessary to adjust the observed data for the effect of scattering by the islands on which the sea-level recording stations were situated. Rather small islands also had to be corrected for the scattering effect. The results showed once more that all sea-level records connected with tsunami waves and made on islands are not representative of the phenomenon in the open oceanic regions.

EFFECT-OF EARTHQUAKES ON SEA LEVEL

Besides the direct effect of the tsunami waves upon the sea level and its fluctuations, earthquakes may cause a considerable rise or subsidence of the sea bottom in the area around their epicentres. The vertical movement of the Earth's crust may influence the mean sea level by changing the zero height of the gauges operating in the coastal regions of the affected area. Yamaguti (1965) paid special attention to the changes in the height of sea level before and after the great Niigata earthquake ($M = 7.5$) which occurred on June 16, 1964. It may be of interest from the standpoint of sea-level research to recapitulate the principal features of the concerned fluctuations.

Measurements of the fundamental elements necessary to prove the accuracy of the standard line in the sea-level records were made on June 14, approximately 45 hours before the occurrence of the Niigata earthquake, in Nezugaseki on the west coast of Hondo island, situated some 40 km from the epicentre of the earthquake. The mean sea level for those 45 hours was determined at 113.3 cm. (It must be mentioned in this connection that the sea level at this station in fact increases when the recorded data show a decreasing trend.) As a consequence of the earthquake, tsunami waves attacked the Bay of Nezugaseki. The mean sea level for the 45 hours extending from about 8 hours after the earthquake was also determined, resulting in an average value of 99.9 cm. The difference between the two values amounts to 13.4 cm and reflects the rise of the mean sea level, or more exactly the subsidence of the sea bottom relative to the mean sea level.

Moreover, the half-monthly mean sea-level heights from June 1 to 15 and from June 17 to 30 were determined by Yamaguti, taking into consideration the perturbating effect of atmospheric pressure and water temperature on the sea level. The departure between these two half-monthly average sea-level heights amounted to 17.7 cm. These numerical examples show that the rise in sea level as a consequence of the Niigata earthquake may be estimated at 15 cm.

Yamaguti also paid attention to the height of the monthly mean sea level at Nezugaseki for 10 months before and 4 months after the occurrence of the Niigata earthquake. The results showed close coincidence with the statistical results based on 103 earthquakes characterized by the magnitude $M \gtrless 7$. The sea-level curve representing the monthly averages reached a maximum 4 months, and a minimum one to 2 months, before the earthquake. In the months immediately following the eruption the mean sea level rose in all the cases which were examined.

Finally, Yamaguti considered in the study the yearly mean sea-level data before the Niigata earthquake. The mean sea level at Nezugaseki started to decrease fairly conspicuously approximately 5 years before the earthquake. Similar features were observed in the case of the Kwanto earthquake, with a magnitude $M = 7.9$ and an epicentral distance of about 40 km, and in connection with the Nankaido earthquake, with a magnitude of $M = 8.1$ and an epicentral distance of approximately 400 km, in the sea-level diagram of yearly means at Aburatubo. These three examples should be sufficient to indicate that there may be some regularities in the changes of the yearly mean sea-level heights which could be utilized for the preliminary prediction of powerful earthquakes in seismic regions.

As another interesting but less-studied example, it may be appropriate to mention that as a consequence of the earthquake in Messina, which occurred on December 28, 1908, the zero of the gauge fell not less than 57 cm and continued to fall, although more slowly, during the following years.

In addition, it may be of interest to mention the results achieved by Hicks (1972a) in connection with a comparison of pre- and post-earthquake sea levels and tidal observations referring to the earthquake in Alaska in 1964. In spite of the fact that the vertical movements of the Earth's crust reached major dimensions, the changes in the tidal harmonic constants were within the limits of normal variability. Also, the non-harmonic tidal characteristics such as mean diurnal high- and low-water inequalities and Greenwich mean high- and low-water intervals did not differ significantly. These results show the pronounced constancy of the Earth in its response to the tide-generating forces.

CHAPTER 9

DETERMINATION OF THE MEAN SEA LEVEL FROM THE RECORDS

In spite of all efforts to supervise the operating processes of a sea-level recording gauge, some more-or-less pronounced and consistent errors in the data can hardly be avoided. The character of these errors depends decisively on the type of gauge, the construction of the gauge well and other contributing factors. It is therefore, rather difficult to formulate a general rule for the occurrence of these errors. Some of the errors may occur continuously, if, for instance, the response of the sea-level variations in the recording well to those in the sea outside the gauge is not satisfactory, owing to inappropriate dimensions — length or diameter — of the pipe connecting the well with the sea. A number of errors may be quasi-periodic; to this group of disturbances belong the effects of mechanical friction, inadequate calibration and hydro-expansion of the chart paper, damping in the records as a consequence of an occasional contraction of the pipe diameter brought about by the penetration into the pipe of sand, gravel or, in higher latitudes, of ice sludge, etc. In addition, there may arise purely random errors, such as unsatisfactory operation, datum instability, etc.

If the scale of the recording gauge is 1/10, the accuracy of the individual readings from the sea-level graph may be estimated at 1.0—1.5 cm, and random errors of this magnitude may be expected to result in standard deviations of 0.6 mm and 0.15 mm in the monthly and annual mean values respectively (Rossiter, 1972a). However, it is by no means excluded that in some cases, in spite of an elaborate control, much larger errors may be inherent in the original data.

There are many different schemes for the computation of the average daily, monthly and annual values from the raw data. The most elementary procedure is the averaging of the 24 daily values, while in other cases sophisticated numerical filters are largely applied. The principal purpose of all these methods is to reduce the effect of the tidal constituents with the frequency of approximately one and two cycles per day. The daily average values may then constitute the starting point for the determination of the monthly and annual mean values. These processes may also contribute to the reduction of the effect of the tidal fluctuations upon the final result.

The contribution of the particular tidal constituents in the case of the elementary averaging of 24-hour sea-level heights to the day is — according to Rossiter (1972a) — given in the following. The values refer to the maximum contribution of the constituents, expressed as percentages of their amplitudes, to the mean values for a 30-day month and a 365-day year. (For a description of the particular tidal constituents see Tables II and IV).

	M_2	N_2	K_1	O_1	M_4	M_6	Msf
30-day month	0.055	0.209	0.267	0.401	0.058	0.060	1.55
365-day year	0.035	0.005	0.000	0.072	0.023	0.008	1.27.

It may be added here, that generally the tide-free mean annual values of sea level may be considered accurate to approaching 1 mm, assuming that the data have been treated with extreme care.

In spite of the fact that the contribution of the tidal constituents may be eliminated from the original data with relatively high accuracy, not only the daily but also the monthly and annual mean sea-level heights show pronounced deviations in time. The following examples may illustrate this fact. In Brest, for the time-span from 1808 to 1964, with gaps covering altogether 23 years, the maximum deviation in mean monthly sea level amounted to 67.0 cm and the maximum difference in average annual sea level to 26.6 cm. Selecting Atlantic City as a representative of the west coast of the Atlantic Ocean, the records for the time-span from 1912 to 1964, for which data are available, show that for the monthly mean sea levels the departures reach the range of 59.7 cm and for the annual average height 25.2 cm. These results may be considered fairly consistent, if we take into account that the recording period for Brest is considerably more prolonged than that for Atlantic City.

The causes bringing about these pronounced deviations are mainly of meteorological origin and the effect of the contributing factors has already been discussed in previous chapter. However, particularly with reference to the annual means, the effect of the long-term changes — land uplift and subsidence and the eustatic factor — must also be taken into account. In addition, abrupt movements of the Earth's crust, caused by seismic or volcanic activity, may appear in some cases as marked anomalies. Since the latter movements, with the exception of pronounced earthquakes, cannot as a rule be observed directly, they may be proved and their effect determined numerically only by means of repeated geodetic precise levellings.

Taking into account all the above-mentioned elements, Rossiter (1960, 1967, 1968, 1972a) introduced regression analysis, based on the equation:

$$Z_Y = \sum_{p=0} a_p Y^p + \sum_{r=1} b_r B_r + c_1 \cos N + c_2 \sin N + \phi_Y$$

where the coefficients a, b, c must be determined with the help of the recorded data. Y is the number of the year — relative, for instance, to 1900; $\sum_{p=0} a_p Y^p$ represents the secular variation given as a polynomial of the order p; B_r is the annual anomaly of the average atmospheric pressure at the station r; $\sum_{n=1} b_r B_r$ corresponds to the meteorological contribution to sea level from both atmospheric pressure and the wind stress and is determined on the basis of triads of stations covering the sea area over which these factors may be considered significant for the sea-level variations; $N = f(Y)$, the mean longitude of the Moon's ascending node; $c_1 \cos N + c_2 \sin N$ represents the nodal tide; and, finally, ϕ_Y covers the summarized contribution to sea level from all remaining sources not mentioned above.

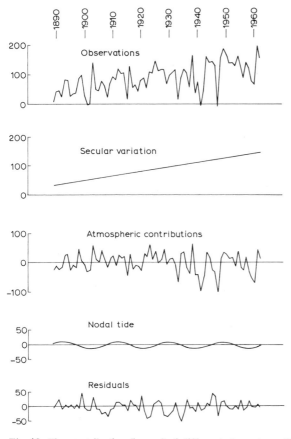

Fig. 40. The contribution (in mm) of different elements to the mean sea level at Esbjerg, 1889–1962 (Rossiter, 1972a).

The advantages of the multivariate analysis of the type given above were illustrated by Rossiter (1972a) for sea-level data from Esbjerg. Fig. 40 shows the various components of the annual anomaly for the years examined. The standard deviations of the original data on sea-level, of the contribution of atmospheric pressure and of the residuals are 49, 35 and 20 mm respectively. The figure indicates a pronounced correlation between the meteorological variable and the mean sea level. The contributing effect of the nodal tide is weak and ill-defined, the amplitude being 9 ± 5 mm. The secular variation shows a linear trend of 1.48 ± 0.15 mm per year. An elementary univariate regression analysis of the data for Esbjerg for the secular variation resulted in a linear trend of 1.21 ± 0.23 mm per year. Although the two trend estimates do not show pronounced deviations, the standard deviation in the former case distinctly indicates that the multivariate analysis is the more preferable of the two methods. It proves also that this analysis should be used to compute not only the annual values of the sea-level heights, but also the average sea level for more prolonged periods.

CHAPTER 10

PRACTICAL ASPECTS OF SEA-LEVEL VARIATIONS

TIDE PREDICTION AND TIDAL TABLES

Before giving a short survey of the present state of tide prediction and its historical background, it may be appropriate to define a few terms connected with the tidal phenomena and used in practical handbooks referring to this subject. Some of the more important terms have already been mentioned in Chapter 1, but in order to have a complete set of the terms they will be repeated in the following compilation.

Spring tides are the semi-diurnal tides within a semi-lunar period of approximately 15 days which have the most pronounced range. They occur near the days of full moon and new moon, when the attraction of Moon and Sun reinforce each other.

Neap tides are the corresponding tides with the least pronounced range. They occur near the times of the first and third quarters of the Moon when Moon and Sun counteract each other.

Throughout a spring–neap cycle the period of the semi-diurnal tides changes. These changes are called the *phase inequality*. In the cases where the period is larger than the average period of the semi-diurnal tidal constituent M_2 (12 h 25 min) the tide is lagging (neap to spring). When the period is less than the average one, the tide is priming (spring to neap).

The main features of the semi-diurnal tide may be defined by 8 characteristics, of which 4 refer to the times of high and low water and 4 to the corresponding heights. Mean high-water interval (MHWI) is the high-water interval (HWI) averaged over not less than one lunation of 29 days, with HWI being the interval between Greenwich Mean Time of the Moon's transit and the standard time of the next high water. Mean low-water interval (MLWI) and low-water interval (LWI) are defined in an analogous way. Mean high-water springs (MHWS) and mean high-water neaps (MHWN) are the sea-level heights of spring and neap water, respectively, averaged over a period covering at least 12 months. Mean low-water springs (MLWS) and mean low-water neaps (MLWN) have a corresponding definition.

Diurnal inequality refers to the differences in the successive ranges and periods of the semi-diurnal tide which is disturbed by the presence of the less pronounced diurnal tide. The term mean higher high water (MHHW), mean lower high water (MLHW), mean higher low water (MHLW) and mean lower low water (MLLW) illustrate a typical diurnal inequality and hardly require any explanation.

Mean sea level (MSL) is the average height of the sea level computed for a given, generally more prolonged, period. Daily, monthly and annual values of mean sea level are as a rule computed from the hourly height data (pp. 205–207). It must especially be pointed out that mean sea level must always be distinguished from *mean tide level* (MTL) for the determination of which the heights of high and low water are used.

The benefit of knowing the times of high and low water in harbours and other shallow regions for navigational purposes has been understood for a long time. According to Rossiter (1972b), the earliest known forecast of high-water time in Britain is credited to John Wallingford, Abbot of St. Albans, who died in 1213. Wallingford assumed the tide at London Bridge to be invariably 3 h 48 min later than the Moon's transit, which itself was considered to be retarded daily by 12 min. From this rather primitive starting point the development has been enormous. Although differing slightly in detail in individual countries the general trend of this development has been everywhere quite similar. It may therefore be sufficient to describe in more detail the development of tidal prediction in one country. The United States has been chosen for this purpose.

The Coast and Geodetic Survey in the United States began work in tidal prediction more than a century ago (Hicks, 1967), the first tide tables being published for the year 1868. They contained low-water predictions for the stations along the Pacific coast and the western coast of Florida. A non-harmonic lunar-tidal interval method was employed at first. This method, to say the least, was extremely cumbersome, involving in its most advanced form the correction of lunar-tidal intervals and heights for the Moon's phase and for the declination and parallax of Moon and Sun. However for the preparation of the tidal tables for the year 1885, the harmonic method was used. This was due to the introduction of a mechanical analogue tide-predicting machine – the Maxima and Minima Tide Predictor – constructed by William Ferrel, which was the very first to be used in the United States. This predictor was designed to cope with the large number of tidal constituents and their summations as required by the method of harmonic analysis. Although only giving the results for 19 constituents, the Ferrel predictor was an improvement over its British predecessors, since the predicted times and heights of high and low water could be read directly from the dial indicators. Low-water prediction for all stations began in 1887.

The Ferrel predicting machine was in operation for a quarter of a century, until Harris and Fischer constructed a predictor covering 37 tidal constituents, which first came into operation in 1912. The Harris–Fischer prediction machine had a height side and a time side. This offered the possibility of having an analogue trace and high- and low-water time and height prediction simultaneously. This machine as such was in operation until 1961, when a motor drive and an automatic readout were added. Hicks should be quoted in this connection: 'Economics, speed and accuracy did not dictate the bow of the mechanic analogue machine to the electronic digital computer until 1966 – a glorious testimony to its efficiency'.

In order to complete the above description of the development of methods for tidal prediction, it may be mentioned that at the suggestion of the Royal Society of London the Hydrographic Office in Great Britain produced its first set of tide tables as early as

1833 (Rossiter, 1972b). These tables gave the predicted times of high water for London Bridge, Sheerness, Portsmouth and Plymouth. By 1870 some 25 British ports appeared annually in the tidal tables. Shortly after the introduction of the harmonic analysis of tidal data (1868), W. Thomson (Lord Kelvin) proposed that a tide-predicting machine should be constructed. Such a machine was built in 1878, several years before the Ferrel predictor, and it allowed the summation of 10 tidal constituents. The first German prediction machine, which was able to sum 20 tidal constituents, was ready for operation in 1916. In 1938 it was replaced by a predictor covering 62 constituents. Constructed according to the specifications by H. Rauschelbach, it has been in operation at the Deutsches Hydrographische Institut in Hamburg. De Rauschelbach predictor made it possible to compute within 20 hours the time of arrival and the heights of high and low water for a harbour for a whole year. In the State Oceanographic Institute in Moscow a tide-predicting machine has been in operation since 1946 which allows the summation of 30 tidal constituents. In the Institute of Coastal Oceanography and Tides in Birkenhead (formerly Liverpool Tidal Institute) a predictor covering 42 tidal constituents started operation in 1950. This machine was constructed under the guidance of the outstanding specialist of tidal phenomena, A. T. Doodson. Photograph 4 shows the imposing outlines of this machine.

A German tide predictor constructed for use on board ship is also available. It is hand operated and covers 10 tidal constituents.

Photograph 4. The imposing outlines of the British mechanic analogue tide prediction machine. (Photograph: Institute of Coastal Oceanography and Tides, Birkenhead.)

There is, moreover, a method of predicting high and low water in a harbour where the tidal observations cover a very restricted time, for instance, only 24 hours. This method may render relatively satisfactory results, if the harbour for which the computations are required is characterized by a fairly regular semi-diurnal tide. To the time of the highest culmination of the Moon is added the time lag and after some minor corrections the time for high water may be determined. The time for low water occurs 6 h 12 min later, always assuming that the tide is regularly semi-diurnal.

A considerable simplification of the work may be achieved by the method of comparing the tidal phenomena in two harbours which have more or less the same character. If for one of these harbours tide tables are already available, corresponding tables are obtainable for the other harbour by means of very simple computations.

The accuracy of tidal prediction is as a rule fairly high. The departures between the predicted and recorded times for high and low water do not generally exceed 10–15 min. Since the sea level, during the occurrence of the extreme water heights, changes very slowly, a time difference of such small magnitude has no practical significance. The accuracy of the predictions may self-evidently be greatly reduced by the contribution of meteorological elements, especially the effects of wind and atmospheric pressure. These effects will be considered more closely in the next section, since this contribution is of the greatest significance for different technical purposes.

For a number of harbours which belong to the regions characterized by shallow water, the deformation of the tidal profile is so accentuated that the method of harmonic analysis does not result in satisfactory data. For these complicated cases other methods must be used. A more detailed and extensive elaboration of these methods is in progress at the present time.

From shallow-water areas the step is not very far to the marginal and Mediterranean-type seas. In the latter sea basins the tides are perceptible, even if generally weak and therefore frequently overshadowed by the contribution to sea-level variations brought about by other factors. Since the practical significance of the tides is slight in these seas, they have, as a rule, not been investigated to a great extent. Nevertheless, a restricted number of harmonic constants for the Baltic Sea, in particular for its large gulfs, are available, and it may be interesting to examine to what extent these constants are representative of the tidal phenomenon in this sea basin. For this purpose the Gulf of Finland has been selected, since the tides are more pronounced in this region than in other parts of the northern Baltic. According to Lisitzin (1944) the greatest possible range of tidal sea-level variations, based on 6 constituents, is 16 cm in the inner part of the gulf. (The corresponding value for Leningrad, situated still farther east in the Bay of Kronstadt, is probably somewhat higher, but unfortunately is not at our disposal.) In the middle part of the Gulf of Finland the largest possible tidal range is according to the computed values 10 cm at Helsinki and 8 cm at Tallinn. Towards the approaches to the gulf the tidal range decreases rapidly, reaching a value of 4 cm.

In order to obtain a concept of the correspondence between the tidal curve based on the computed harmonic constants and the actual records, it was necessary to find a time-span for which the fluctuations in sea level caused by the meteorological factors were not marked. In this connection it may be pointed out that these fluctuations are very accentuated in the inner parts of the Gulf of Finland, reaching at Koivisto, for instance, an amplitude of more than 280 cm (Stenij and Hela, 1947). A period corresponding to weak sea-level variations cannot therefore be long. However a time-span of a few days is assuredly sufficient to depict the characteristic features of the particular stations. The period selected for the purpose covered 6 days, May 6–11, 1936. In Fig. 41 are reproduced the recorded and the computed tidal curves for 6 sea-level stations in the Gulf of Finland. The recorded values were reduced to the average sea level, based on consecutive means for 25 hours. The largest deviation of these mean values did not exceed 10 cm,

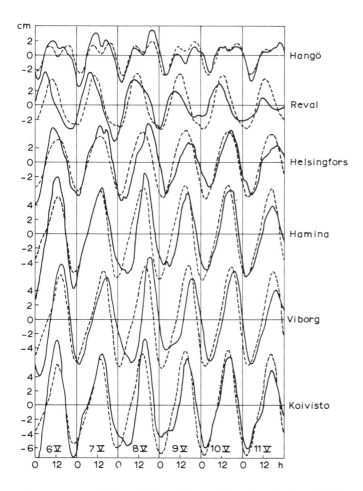

Fig. 41. Tidal variations at the stations in the Gulf of Finland during the period May 6–11, 1936. Solid lines refer to the recorded sea level, dashed lines to the computed sea level (Lisitzin, 1944).

which must be considered a satisfactory result. The computed sea level is based on the harmonic constants, which for Tallinn (in the figure denoted as Reval), Helsinki and Koivisto cover a period of 8 years and for the 3 remaining stations of only 4 years. The figure shows that the tides are mixed at Hangö (Hanko), but have pronounced diurnal characteristics for the other stations. The coincidence between the observed and computed curves must be considered fairly satisfactory. The departures are at their greatest on May 8, and are especially marked concerning the minimum heights. During the three following days also there is a distinct displacement of the curves. For the stations situated on the northern coast of the Gulf of Finland the extreme values of the computed curves occur earlier than the recorded ones, while for Tallinn on the southern coast the computed maximum and minimum heights are delayed in comparison with the corresponding observed values. Concerning the amplitudes it may be mentioned that the recorded curves show at the beginning of the research period a more pronounced range than the computed curves, while the situation seems to be in reverse towards the end of the period. These results indicate that for sea basins with weakly pronounced tides a prediction, or, to be more exact in this case, a post-determination of the phenomenon is also possible, although only more-or-less approximate values may be obtained.

TECHNICAL ASPECTS AND COAST PROTECTION – SEA-LEVEL STATISTICS

In addition to the secular variation in sea level the heights of highest and lowest water are important factors in the design and construction of harbours and other coastal installations. In areas where the tidal variations are pronounced even a minor addition to the extreme heights may be significant, while in shallow Mediterranean-type seas with weak tides the effect of the meteorological elements may be so accentuated that accurate knowledge of the extreme range of sea-level is a 'must' for coastal engineering.

Duvanin (1956) contributed some interesting views and results to the problem of the combined effect of tides and hydro-meteorological factors. According to Duvanin, attention must be paid to the fact that the extreme departures of sea level from the average value may not be considered as the sum of two contributing factors. The most pronounced changes in sea level depend upon three factors: the range of the tide, the range of sea-level variations caused by hydro-meteorological factors and the frequency of the extreme fluctuations in sea level. Knowledge of the frequency is very important, since the tidal variations on the one hand and the hydro-meteorological increases and decreases in sea level on the other hand show a very complex picture, following quite different laws. For the largest departures, which are the result of interaction of the two factors, the probability will decrease rapidly with increasing range.

On the basis of prolonged series of sea-level observations in regions with marked tides, characterized by well-developed semi-monthly deviations, it can be established that the extreme high- and low-water heights are generally the consequence of the tidal variation. The contribution of sea-level fluctuations, due to hydro-meteorological factors, to the extreme heights increases in magnitude with the decrease of the range of the tides. It is, of course, self-evident that in sea regions with practically no tidal variations the hydro-meteorological contribution is reponsible for all sea-level changes.

In order to give the characteristics of the regime of sea-level variations as a consequence of tides and hydro-meteorological factors, two terms h_t and h_Σ were introduced by Duvanin. h_t represents the highest range of sea-level changes due to the tides and is determined on the basis of the harmonic tidal constituents. h_Σ corresponds to the most pronounced range, recorded for the same station. The ratio h_t/h_Σ varies within the limits of zero and one. In coastal regions where the tides are very weak the ratio is close to zero, while in areas with pronounced tidal variations the ratio tends to approach the value one.

The coincidence of the occurrence of tidal high water and the peak of a storm surge at Southend in the Thames estuary was studied by Dines (1929). The authors classified 37 well-developed storm-surge cases according to the time of their occurrence at Southend in relation to the tidal high water. Dines' results are given in the following:

	Hours before high water						Hours after high water					
	5–6	4–5	3–4	2–3	1–2	0–1	0–1	1–2	2–3	3–4	4–5	5–6
Frequency	0	6	14	6	2	0	1	2	2	1	3	0

These frequencies refer to cases where the sea level exceeded the predicted tidal heights by 120 cm or more. The values indicate that storm surges show a tendency to occur at half-tide and to avoid the time of high tide. It is thus interesting to note that only in one case was the peak of the storm surge observed within one hour of high water and in five cases within two hours of high water. In one of these cases the increase in sea-level above the predicted height was 150 cm. This occurred in January 1928, causing a flood of disastrous proportions. The tendency of the storm surges at Southend to occur during the period of increasing tide is probably a consequence of the interaction between the meteorological phenomenon and the astronomical tide. A considerable amount of work has been devoted to the problem of this interaction, examining theoretically the simultaneous motion of tide and storm surge in a narrow channel or estuary, but the phenomenon as a whole is still not completely understood. Nevertheless, reference may be made to the studies performed on this interesting problem by Proudman (1955a, 1957, 1958), Doodson (1956), Rossiter (1961) and Rossiter and Lennon (1965).

Another example of the interaction of the meteorological contribution to sea level with extreme values of the tides was given by Duvanin. Using sea-level data for the island Sosnovets, recorded during a period of 80 months, elaboration of the results showed that the probability of the occurrence of the different sea-level heights due to the piling-up effect of the wind was as follows:

Probability (percent)	99	98	95	90	85	75	25	15	10	5	2	1
Sea level (cm)	−41	−36	−29	−23	−19	−13	10	16	21	30	40	42

TABLE XLIII

THE TIDAL RANGE AND THE TOTAL OBSERVED SEA-LEVEL VARIATIONS

Station	Years of observation	h_t (cm)	h_Σ (cm)	$\bar{h}_t$ (cm)	$h_t - \bar{h}_t$ (cm)	h_t/h_Σ
Baltimore	25	57	318	41	16	0.18
Solombala	16	83	269	52	31	0.31
Washington	7	133	404	94	39	0.33
Kem	13	193	314	118	75	0.61
Basra	6	195	310	123	72	0.63
Philadelphia	20	223	411	158	65	0.54
Ekaterinskaja Gavan	10	407	496	240	167	0.82
Boston	10	423	564	297	126	0.75
Sosnovets	7	475	495	304	171	0.90
Ketchikan	6	740	764	480	260	0.97

The piling-up effect is thus quite evident in the sea-level variations at Sosnovets, the range of this effect being 83 cm. However, the greatest observed range of sea-level variations which has been noted at Sosnovets during the recording period was only 20 cm larger than the most pronounced range of the tides. This shows very distinctly that during the period concerned the marked tides did not coincide with the most accentuated meteorological contribution.

Duvanin has, moreover, extended his research concerning the greatest tidal range and the most pronounced total sea-level range to cover more extensive data. For this purpose he introduced, in addition to the ratio h_t/h_Σ, a term representing the difference between h_t and $\bar{h}_t$, where $\bar{h}_t$ corresponds to the average range of the tide.

The results obtained in this connection by Duvanin are reproduced in Table XLIII, which refers to 10 different sea-level stations situated in different parts of the world's oceans and representing highly deviating tidal types.

The data in Table XLIII are presented graphically in Fig. 42. In this figure the ratio h_t/h_Σ represents the ordinate and the difference $h_t - \bar{h}_t$ the abscissa. A characteristic feature of this figure is that the position of the points, as a result of the accumulation of new and more precise values for h_Σ, are only likely to move downwards. The concerned ratio will reach its minimum value in the case where h_Σ becomes equal to $h_t + h_m$, h_m corresponding to the largest possible variation in sea level due to the hydro-meteorological contribution. Using the values given in Table XLIII, the hyperbola:

$$\frac{h_t}{h_\Sigma} = \frac{h_t - \bar{h}_t}{87 + 0.7(h_t - \bar{h}_t)}$$

was determined by Duvanin. It is represented by the solid curve in Fig. 42. According to Table XLIII, the conclusion may be drawn that, on average:

$$\bar{h}_t = 0.65\, h_t$$

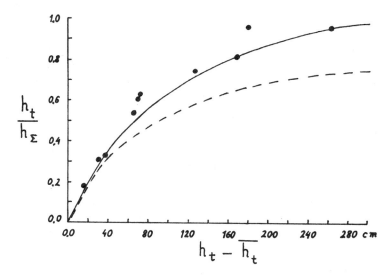

Fig. 42. The relationship between the terms h_t/h_Σ and $(h_t - \bar{h}_t)$, where h_t is the maximum range of the tide, $\bar{h}_t$ the average range of the tide and h_Σ the maximum observed range of sea-level variations caused by the added effect of tides and the hydro-meteorological contribution (Duvanin, 1956).

Taking into account this relationship, we obtain:

$$\frac{h_t}{h_\Sigma} = \frac{h_t}{250 + 0.7\,h_t}$$

The value 250 in this formula corresponds to the average maximum range of the meteorological contribution (in cm) determined on the basis of the data in Table XLIII. With the help of this value the dashed curve in Fig. 42, representing the minimum of the ratio h_t/h_Σ, was computed. A comparison of the two curves shows that the probability of the coincidence between the maximum range of the tide and the hydro-meteorological effect decreases with the increasing range of the tide.

In this connection it must be pointed out that the above results are based on a rather restricted amount of sea-level data. Duvanin himself paid attention to the fact that the periods used for the computations are in some cases too limited to allow definite conclusions to be made. If the results had been based on data covering, for instance, 100 years, the hyperbola represented by the solid curve would have been situated somewhat lower. It is also evident that the results obtained by Duvanin are not generally valid, since local conditions, and in particular the depth of the water in the vicinity of the tidal stations, have not been taken into consideration.

Disney (1955) has to some extent considered the same problem as Duvanin, having at his disposal all available tidal data from the United States. Since Disney gave not the maximum range of the tide, but the average yearly maximum range, the data are not quite comparable with those given by Duvanin. Disney did not attempt to achieve generalized results, but paid more attention to local conditions and special cases. Thus Disney mentioned that for localities such as Key West (Florida) where the movement of the water masses is obstructed very little, the sea-level heights are only weakly affected by storm surges. Position features such as those characteristic of Providence (Rhode Island) have as a consequence a funnel or pocket effect which amplifies the height of the sea level. The maximum sea level recorded along the coasts of the United States occurred, according to Disney, at Providence on September 21, 1938, at a height of 4.7 m above mean high water. This increase was the result of a powerful tropical storm that caused much damage in the affected region. In addition, the data given by Disney showed that this storm also caused extremely high sea levels at a number of other localities, extending from Woods Hole (Massachusetts) to Willets Point (New York). According to the data given by Disney the highest recorded sea-level height in the Gulf of Mexico occurred at Galveston (Texas) on August 16–17, 1915, this height being roughly 3 m above mean high water. Along the Pacific coasts of the United States the maximum sea-level heights are less pronounced. The most marked cases, for Skagway and Anchorage (Alaska), have shown an elevation of 1.75 m above the mean higher high water.

A very interesting and comprehensive investigation of the frequency of abnormal high sea levels along the west coast of Great Britain was completed by Lennon (1963). This oceanographer based his study on different statistical methods in order to extrapolate the trends of the frequencies necessary for determining the optimum height for sea-coast protection. The results of the particular techniques were used to estimate the height of the sea level which may be reached once in one hundred years and once in two hundred years.

The records which were selected by Lennon for his study varied to some extent for the different tidal stations, but the lower limit of the sea-level data was, on average, fixed at approximately 30 cm above Mean High-Water Springs (MHWS). The results for Liverpool, the locality where the records covered the longest span of time — 61.5 years — are presented in Table XLIV. In this table H is the height in cm (transformed from the heights in feet, given by Lennon) above Ordnance Datum (Newlyn), n is the number of cases during the period of tidal records for which the height H has been attained or exceeded. N is the total length of the period of tidal records used in the study, expressed in years and adjusted for breaks. The ratio n/N corresponds thus to the average number of cases per year for which the height H has been attained or exceeded. Finally, h is the height in cm above the level of MHWS, as given in the Admiralty Tide Tables, 1960.

If a logarithmic scale is used for the average number of cases per year (n/N) and

TABLE XLIV

THE FREQUENCY OF HIGH SEA LEVELS AT LIVERPOOL DURING 61.5 YEARS

H (cm)	473	488	503	518	534	549	564	579	595	610	625
n	2244	1293	663	313	123	39	13	5	3	1	0
n/N	36.5	21.0	10.8	5.1	2.0	0.63	0.21	0.08	0.049	0.016	0
h (cm)	·40	55	70	85	101	116	131	146	162	177	192

plotted against the height H above the Ordnance Datum, the result for Liverpool is a linear presentation of the plots, which indicates that there exists a simple basis for extrapolation. Concerning many of the other localities investigated by Lennon the plotted data show a distinct curvature. This fact is by no means surprising, since when studying the phenomenon it must always be borne in mind that the interaction between the effect of atmospheric pressure, wind stress and Coriolis force, on the one hand, and the tidal motion, on the other hand, is very complex. In the cases where the sea level is only 30–50 cm above MHWS, the concerned heights may be reached by the ordinary tide without contribution from the meteorological factors. For somewhat more pronounced sea-level heights the range may depend on the added effect of the tidal regime and the meteorological contribution, while for the most extreme cases the effect of the wind-produced piling-up of the water may be the predominant factor. Therefore, an assumption seems to be reasonable that even in localities showing a curvature for the lower sea-level heights, the highest levels tend to approach the linear conditions.

For Liverpool, where the relationship between the height and the logarithmic scale of the ratio n/N was distinctly linear, the following equations were deduced by Lennon (here converted into cm):

$$\text{Log}_{10} (n/N) = -0.0247\,H + 13.367$$
$$\text{Log}_{10} (n/N) = -0.0247\,h + 2.674$$

On the basis of the above equations it can be computed that the probable maximum sea level to be reached in Liverpool once in 100 years in 622 cm above Ordnance Datum, and the level to be attained once in 200 years in 634 cm above the same datum.

Since the local authorities in a number of ports in Great Britain compare daily the observed and predicted heights of sea level, the deviations between the two series of data may be considered as a measure for the contribution of the meteorological factors. Self-evidently it must be assumed in this connection that the predicted tide is fully reliable. This kind of information being available for Liverpool, it is possible to examine the occurrence of storm surges and their effect at the time of high water in this port.

If S denotes the positive storm-surge contribution to the height of high water, determined as the difference between the observed and the predicted sea-level heights, n^s the number of cases during the period of the records for which the surge contributions have been exceeded and n^s/N the average number of cases per year of the surge contribution, the results determined by Lennon for the records at Liverpool for the time-span of 17.6 years are given in Table XLV.

TABLE XLV

THE FREQUENCY OF THE POSITIVE STORM-SURGE CONTRIBUTION TO SEA-LEVEL HEIGHTS OF HIGH WATER AT LIVERPOOL

S (cm)	46	61	76	92	107
n^s	149	54	14	4	0
n^s/N	8.43	3.06	0.79	0.23	0

It is now possible to analyze the data in the manner previously described for abnormally high sea levels. The equation:

$$Log_{10}(n^s/N) = -0.034 S + 2.51$$

was given by Lennon. On the basis of this equation the probable storm-surge contribution to high water to be exceeded once in 200 years at Liverpool is 141 cm.

In addition to the logarithmic function, Lennon also considered other statistical solutions of the problem. Since most of the investigations of the frequency are concerned with a bell-shaped distribution of the cases with frequencies declining on either side of a central maximum and, owing to the markedly physical character of the problem, it was impossible to take into account the complete distribution of all high-water levels, it seemed reasonable to obtain the bell-shaped form by reducing the data and presenting them as annual maximum levels. It is an accepted procedure to plot the data according to a compromise frequency and two such systems are in common use. According to the system introduced by Gumbel the expression $m/(n+1)$ is used, where m denotes the rank of the record and n the total number of records involved in the computations. For different technical purposes the system given by Hazen is more commonly adopted. This system is based on the expression $(2m-1)/2n$. As a rule, the deviations between the final results of the two alternatives are only slight. For the computations the annual maximum heights are arranged in order of the rank (m), and the expressions $m/(n+1)$ and $(2m-1)/2n$ are determined and converted into a measure of probability by writing, for instance, for the system proposed by Hazen:

$$y = -Log_e Log_e \frac{2n}{2m-1}$$

According to Gumbel the probable occurrence of the sea-level height H above the Ordnance Datum is given by an equation of the form:

$$H = D + E y,$$

where, if F is the frequency of the occurrence of a sea-level height in years:

$$y = -Log_e Log_e [F/(F-1)]$$

According to the results of this equation the probable maximum sea level to be attained at Liverpool in 100 years is 625 cm above Ordnance Datum, the corresponding height for 200 years being 640 cm.

Lennon considered in his paper four additional procedures for the determination of the highest probable sea level to be reached once during one and during two centuries. These procedures showed that the results are fairly consistent, varying for the period of 100 years between the limits of 597 cm and 625 cm, and for the period of 200 years between the limits of 604 cm and 640 cm. The average value achieved on the basis of the six methods used by Lennon was 611 cm for 100 years and 621 cm for 200 years.

Of the ports situated on the west coast of Great Britain Avonmouth is the most exposed, with a probable maximum sea level of 881 cm to be attained once in 100 years and a sea level of 906 cm to be reached once in 200 years, both heights given above the Ordnance Datum.

Lennon gave in his paper an interesting synthesis of the significance of the probable 100-year maximum sea level, by determining the amount by which this maximum height exceeds the height of MHWS in relation to the spring range of the tide in the same port. The following results were achieved by Lennon:

	Difference between 0-year maximum and MHWS in relation to spring range
Avonmouth	0.14
Newport	0.16
Cardiff	0.14
Swansea	0.16
Llanelly	0.23
Liverpool	0.21
Preston	0.30
Fleetwood	0.19
Heysham	0.21
Silloth	0.23
Dublin	0.28
Belfast	0.43.

The above values indicate that in the highest reaches of the Bristol Channel the probable 100-year maximum sea level exceeds MHSW by 0.15 times the spring range. Although storm surges are known to develop within the channel in the same manner as the incoming tide, it is evident that the local exposure towards the southwest may contribute somewhat to the increase in sea level. Newport and Swansea are slightly more exposed than Avonmouth and Cardiff, with the result that the ratio is somewhat higher. It is by no means excluded that in the northern part of the channel the ratio may increase to 0.17, and conversely along the southern coast of the channel the reduction of the ratio to 0.13 is possible. In the western part of the channel the ratio will most probably increase as a consequence of the decreasing tidal range to roughly 0.23, as is indicated by the results for Llanelly. In the Irish Sea an average ratio of 0.21 may be expected, taking into account, however, that the value for Preston is less reliable. An average value of 0.3 may be assumed for the west coast of Great Britain, this value increasing towards the north as a consequence of the reduction in tidal range.

As a contribution to the study of the significant problem of the interaction between tide and storm surge, Lennon completed his results by a number of tables illustrating this phenomenon. Since from theoretical considerations it seems to be probable that the meteorological contribution will be more pronounced in shallow than in deep water, the timing of the onset of the storm-surge effect in relation to the height of the tidal wave may be of interest. The depth of the water participating in the development of the storm surge also has theoretical significance. Lennon therefore investigated the evidence for the possible interaction between tide and storm surge on the basis of the data at his disposal. The results for Liverpool are given in Table XLVI, converting the data from feet into centimetres.

Table XLVI shows that the distribution of the predicted heights of high water and also of the storm-surge contribution is asymmetrical, the former being weighted towards the higher levels, the latter towards the lower levels. The probability of a more marked storm-surge contribution to high water at neaps is approximately 25% higher than at springs. A meteorological contribution exceeding 45 cm covers only 11.5% of all high-water cases.

For technical purposes, especially in shallow sea regions, knowledge of the frequency and duration of different sea-level heights may be of considerable importance. This knowledge is based on statistics obtained from recorded data covering a prolonged span of time. Experience in Finland has shown that a period of 10 years is generally sufficient to achieve satisfactory results. Exceptions are, of course, frequencies characterizing the most extreme cases, since these results are based on comparatively few data.

In order to give a picture of the distribution of sea-level heights during a given period, e.g., one year, the navigation period (in areas with a seasonal ice cover), the vegetation period or a special month, the data must be arranged according to decreasing levels. The next step is to divide the sequence into groups corresponding to fixed height intervals, let us say, one or five cm. Thereafter the number of cases for each interval is determined and expressed as a percentage of the total data for the concerned period. It hardly needs to be pointed out that the percentage reaches the highest values for medium-height sea levels and diminishes rapidly with increasing and decreasing heights.

Using the percentages characterizing the frequency of different sea-level heights the safety percentage or probability of levels may be determined. In this case the frequency

TABLE XLVI

THE DISTRIBUTION AND GROUPING OF POSITIVE STORM SURGES AGAINST THE PRE-DICTED HEIGHTS OF HIGH WATER AT LIVERPOOL DURING THE YEARS 1941–1958*

Predicted heights (cm) of high water above chart datum	580– 639	640– 700	701– 761	762– 822	823– 883	884– 944	945– 1005
The distribution of the predicted heights of HW (cm)	97	1465	2716	3212	3270	1517	190
Storm surges exceeding 45 cm at HW	1	36	63	54	37	28	2
Storm surges exceeding 45 cm at HW in percentage of HW cases	1.05	2.5	2.3	1.7	1.1	1.8	1.05

* Chart Datum is 453 cm below Ordnance Datum, MHWN = 695 cm, MHWS = 876 cm.

percentages are continuously added, beginning with the highest sea level and ending with the lowest. The level representing the safety percentage of 50 is generally rather close to the mean sea level, but may in some cases show a more-or-less pronounced deviation. It is called the medium level. The term safety percentage is mainly used by Russian oceanographers. In cases where no confusion may arise, the term 'frequency' is also used in the following instead of safety percentage.

TABLE XLVII

THE SEASONAL VARIATION OF THE FREQUENCY OF SEA-LEVEL HEIGHTS (CM) AT HEL–SINKI, PERIOD 1926–1955

%	J	F	M	A	M	J	J	A	S	O	N	D
Max.	108	87	95	89	48	54	51	66	84	112	104	98
0.1	93	79	70	69	41	44	49	60	71	83	83	86
0.2	84	75	66	63	36	42	47	56	67	79	79	81
0.3	79	71	62	60	33	41	46	54	64	75	75	78
0.4	73	68	60	57	32	40	44	52	62	73	73	76
0.5	70	66	56	55	31	39	43	51	60	72	71	74
1	63	59	46	48	27	35	39	46	55	65	64	65
2	57	54	39	40	24	31	34	41	49	59	55	59
3	52	50	34	34	22	28	31	38	46	55	50	55
4	49	47	31	31	20	26	30	36	43	52	47	50
5	46	44	29	29	18	24	28	34	41	49	44	48
10	38	33	19	20	11	17	22	29	34	41	35	40
15	31	25	12	15	5	13	19	24	28	36	29	34
20	26	19	8	10	0	10	16	21	25	32	24	29
25	21	15	3	7	−3	7	14	18	21	28	21	25
30	17	11	0	3	−6	5	12	16	18	24	17	21
40	9	3	−6	−2	−11	0	9	11	13	18	11	14
50	2	−4	−11	−8	−14	−4	6	6	8	11	5	8
60	−5	−13	−17	−13	−17	−7	3	2	3	5	−1	2
70	−11	−24	−22	−18	−21	−10	1	−2	−2	−1	−7	−5
75	−15	−29	−24	−21	−23	−11	−2	−4	−5	−5	−10	−9
80	−19	−33	−27	−24	−25	−14	−4	−6	−8	−8	−14	−12
85	−24	−39	−31	−28	−27	−16	−6	−9	−11	−13	−18	−17
90	−30	−45	−36	−32	−30	−20	−9	−12	−16	−18	−24	−23
95	−39	−52	−43	−38	−34	−24	−13	−18	−22	−25	−33	−32
96	−42	−54	−45	−40	−36	−26	−14	−19	−23	−27	−36	−35
97	−46	−56	−47	−42	−38	−27	−15	−21	−26	−31	−38	−37
98	−52	−60	−50	−45	−40	−29	−17	−23	−28	−35	−41	−40
99	−59	−64	−56	−49	−44	−32	−19	−25	−32	−42	−47	−47
99.5	−65	−71	−61	−52	−47	−35	−21	−28	−35	−48	−52	−55
99.6	−67	−73	−62	−53	−48	−37	−22	−28	−35	−49	−53	−56
99.7	−69	−75	−63	−54	−49	−38	−23	−29	−36	−51	−54	−58
99.8	−72	−78	−64	−55	−50	−41	−24	−30	−37	−54	−57	−61
99.9	−77	−79	−65	−59	−53	−44	−27	−31	−39	−58	−61	−67
Min.	−88	−79	−78	−88	−57	−50	−35	−37	−42	−66	−71	−80

In regions where there are no pronounced vertical movements of the Earth's crust the computations of the frequency distribution are simple, since all data are comparable. On the contrary, in regions characterized by land uplift, for example, the apparent height of the sea level is gradually decreasing and the determination has to be made separately for each year. As a starting point for the computations the average sea level for the whole period under consideration may be chosen, and the annual means for individual years determined using the rate of land uplift. In this way theoretical sea-level averages are determined. In this connection it may be of interest to mention that these theoretically computed sea-level averages may deviate from the actually recorded annual mean sea-levels by amounts reaching 15 cm. Deviations of this magnitude in both directions have for instance been noted for the sea-level stations situated in the Baltic.

In order to illustrate the seasonal cycle of the frequency, or safety, percentage distribution of sea-level for a selected station, Table XLVII has been reproduced from a publication giving the frequencies of sea level along the Finnish coast (Lisitzin, 1959b). This table refers to the conditions at Helsinki and shows distinctly the marked difference between the calm spring and summer months on the one hand and the stormy autumn and winter months on the other hand. As a general rule for the Finnish coastal area, it may be said that the range of sea level changes in July averages roughly half of the range in December and January.

Table XLVIII gives the yearly frequencies of sea-level heights for all the Finnish sea-level stations which were in operation during the 30-year period 1926–1955. This table contains, in addition, not only the recorded maximum and minimum heights of sea level but also the mean high- and mean low-water levels (MHW and MLW). The table shows an accentuated decrease in the frequencies of high and low sea-level heights from the north (Kemi) and the east (Hamina) towards the southwest (Degerby). This is a consequence of the piling-up effect of water caused by wind in the inner parts of the Gulfs of Bothnia and Finland (cf. Fig. 46).

The considerable influence of the position of sea-level stations and their surroundings on the frequency of different sea-level heights is distinctly demonstrated in Table XLVIII. For instance, the frequency distribution at Turku is closest to that at Mäntyluoto, although the latter station is situated in the Gulf of Bothnia and roughly 120 km to the north of Turku. The effect of the extensive archipelago off Turku is reflected in this frequency distribution.

Table XLVIII indicates that the sea-level heights corresponding to a given percentage – or frequency or probability – have a fairly regular course along the coast. A consequence of this regularity is that the table, and similar tables elaborated for other coastal regions according to the same principles, may also be used for the determination of the frequency of the particular sea levels for localities situated in between the sea-level stations. As a rule, a linear interpolation gives satisfactory results; in some cases attention must also be paid to the specific position of the locality concerned in relation to the configuration of the coast and the presence of islands.

Knowledge of the frequencies of different sea-level heights makes it possible to answer

TABLE XLVIII

THE AVERAGE YEARLY FREQUENCIES OF SEA-LEVEL HEIGHTS (CM) ALONG THE FINN-
ISH COAST, PERIOD 1926–1955

%	Ke-mi	Ou-lu	Raa-he	Pie-tar-saa-ri	Vaa-sa	Kas-ki-nen	Män-ty-luo-to	De-ger-by	Tur-ku	Han-ko	Hel-sin-ki	Ha-mi-na
Max.	160	144	124	111	105	93	94	81	100	90	112	139
MHW	118	108	97	79	74	69	69	57	70	63	82	105
0.1	113	102	95	83	75	70	70	59	70	65	79	100
0.2	105	93	85	76	69	65	64	55	64	60	73	88
0.3	97	89	81	72	65	62	60	53	60	58	69	83
0.4	92	85	79	70	63	60	58	52	58	55	66	79
0.5	89	82	76	68	60	58	56	50	56	53	63	76
1	78	72	68	60	54	52	51	46	50	48	57	67
2	67	64	58	53	48	46	45	39	43	44	49	58
3	60	56	52	48	43	42	41	36	40	40	45	53
4	55	52	48	44	40	40	38	34	37	37	42	49
5	51	48	45	41	38	37	36	33	35	35	39	45
10	38	36	34	32	30	29	29	26	27	27	29	34
15	30	28	27	25	24	23	23	21	22	21	24	27
20	23	22	21	20	19	19	19	17	18	17	19	22
25	18	18	17	16	16	15	15	14	15	13	16	17
30	14	15	13	13	12	12	12	11	12	10	12	13
40	7	7	7	7	7	6	7	6	6	5	7	6
50	0	1	0	1	2	1	2	1	1	0	0	0
60	−6	−5	−5	−5	−4	−5	−4	−4	−4	−5	−5	−6
70	−13	−12	−12	−11	−10	−10	−10	−9	−10	−11	−11	−12
75	−17	−15	−15	−14	−13	−13	−12	−13	−13	−14	−14	−15
80	−21	−20	−20	−18	−17	−17	−16	−16	−17	−18	−18	−19
85	−26	−25	−24	−22	−22	−21	−20	−20	−21	−21	−22	−24
90	−32	−31	−30	−28	−27	−26	−25	−25	−26	−26	−27	−30
95	−42	−40	−40	−37	−35	−34	−33	−32	−33	−33	−34	−38
96	−45	−42	−43	−40	−37	−37	−35	−33	−35	−35	−37	−41
97	−49	−46	−46	−43	−40	−40	−37	−35	−37	−37	−40	−45
98	−54	−51	−49	−46	−44	−43	−40	−39	−41	−40	−44	−50
99	−62	−57	−57	−52	−49	−46	−45	−42	−47	−47	−51	−57
99.5	−71	−65	−64	−57	−54	−51	−50	−49	−55	−55	−57	−64
99.6	−74	−67	−67	−59	−55	−52	−51	−50	−56	−56	−58	−66
99.7	−78	−71	−71	−61	−57	−54	−52	−51	−58	−57	−60	−69
99.8	−86	−78	−78	−65	−60	−57	−54	−53	−60	−60	−62	−72
99.9	−97	−90	−88	−72	−64	−60	−58	−57	−63	−64	−67	−78
MLW	−91	−87	−84	−72	−65	−57	−54	−44	−50	−53	−65	−81
Min.	−134	−130	−129	−113	−100	−87	−80	−72	−75	−80	−88	−109

TABLE XLIX

THE MAXIMUM RECORDED (A), THE HIGHEST DAILY (B), THE HIGHEST MONTHLY (C), THE AVERAGE MONTHLY (D), THE LOWEST MONTHLY (E), THE LOWEST DAILY (F) AND THE MINIMUM RECORDED (G) SEA-LEVEL HEIGHTS (CM) IN HELSINKI DURING THE 50-YEAR PERIOD 1904–1953

	J	F	M	A	M	J	J	A	S	O	N	D	Mean
A	108	88	117	89	54	54	78	76	94	100	119	115	92
B	86	66	64	66	36	46	45	60	58	72	79	100	65
C	42	34	29	30	9	18	29	30	31	39	43	53	32
D	4	−2	−11	−10	−13	−3	5	8	8	5	2	7	0
E	−38	−51	−42	−40	−40	−20	−14	−15	−23	−33	−27	−31	−31
F	−76	−76	−83	−75	−54	−44	−29	−34	−36	−55	−62	−70	−58
G	−93	−83	−94	−88	−69	−59	−34	−50	−42	−80	−72	−81	−70

a great number of technical questions. Experience, however, has shown that in particular the annual course of sea-level variations requires some additional investigations. For instance, it may be necessary to know not only how frequently a sea-level height may occur, but also for how long it may last. To some extent this question is answered by Table XLIX, which was computed for Helsinki on the basis of sea-level records for the 50-year period 1904–1953 (Lisitzin, 1954a). This table compared with Table XLVII shows the dependence of the maximum and minimum recorded sea levels upon the period.

Table XLIX also indicates very distinctly the seasonal cycle of sea-level behaviour at Helsinki. For instance, the difference in height between the highest and lowest monthly sea level is in June only 38 cm and in December 84 cm. It may also be noted that the mean height difference between the maximum recorded and the highest daily sea-level heights amounts to 27 cm, being between the minimum recorded and the lowest daily sea levels only 12 cm. The marked deviation in height differences distinctly proves the fact that the maximum sea-level heights are not only more pronounced — as a result of the wind-produced piling-up effect on the water surface — but also of considerably shorter duration. Attention must also be paid to the fact that the departures of the highest monthly and the lowest monthly sea-level heights are on average of the same magnitude, amounting tot 32 cm and −31 cm.

In order to give a graphical illustration of the results in Table XLVII, these are reproduced in Fig. 43.

Since the frequency of the extreme heights of sea level is of the greatest interest and above all of greatest technical significance for a number of coastal constructions and other works, special attention has been paid to these cases (Lisitzin, 1957c). The limit for the extreme sea-level heights has been chosen as the height of 50 cm above and below mean sea level of the year. Helsinki was again one of the sea-level stations investigated in this respect and the time-span was 1904–1953. The research followed three different lines: the determination of the average number of hours during which a given high or low

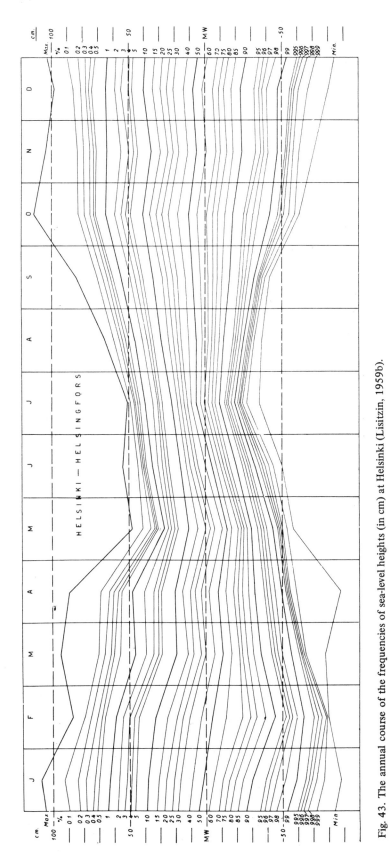

Fig. 43. The annual course of the frequencies of sea-level heights (in cm) at Helsinki (Lisitzin, 1959b).

height of sea level had been reached or passed; the determination of the average duration of high and low sea levels; and, finally, determination of the maximum duration of high and low sea-level heights.

The first-mentioned task corresponds, of course, to the determination of the frequencies already reproduced in Table XLVII, The computed data differ from the results of this table in so far as they refer to another, more prolonged period of recorded data and are given not in percentages, but as average number of hours per year. They are presented graphically in Fig. 44 and 45 for the height group of 50–85 cm concerning high sea levels and for the height groups of −50 to −75 cm concerning low sea levels. For still more pronounced sea-level departures the occurrence is so weak that a graphical presentation is hardly appropriate. Supplementary data are therefore given in Tables L and LI. Since the

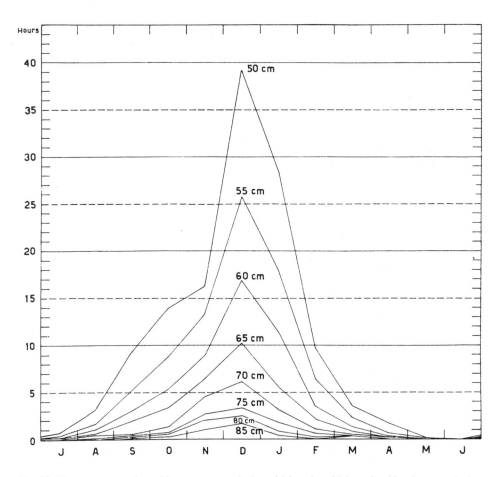

Fig. 44. The average number of hours per year during which a given high sea level has been reached or passed at Helsinki (Lisitzin, 1957c).

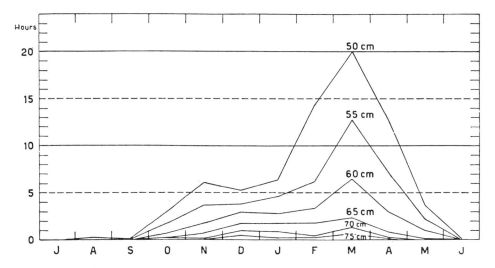

Fig. 45. The average number of hours per year during which a given low sea level has been reached or passed at Helsinki (Lisitzin, 1957c).

occurrence of high sea levels is concentrated around the turn of the year, it seemed more convenient to have the year running from July to June, in order to avoid splitting the tops of the curves in Fig. 44 and 45.

Table L gives the average number of hours per year during which a high sea level has been reached or passed. Since the two months December and January show the most pronounced frequencies of high sea levels, the corresponding numbers are reproduced as such and as percentages of all the cases. The table shows very distinctly that practically half of all cases characterized by a high sea level occur during two months of the year, i.e., December and January.

Concerning the low sea levels, the concentration of data is typically of three months of the year, i.e., Feburary, March and April. Table LI is therefore compiled in such a manner as to accentuate this distribution. It may be noted that more than half of the low sea-level heights occur during this period.

In addition to the frequency of the extreme values of sea level, the duration is an important factor. Attention to this factor is paid in Table LII–LV. The first of these tables gives the average duration of high sea levels at Helsinki, the second that of the low sea levels, while Table LIV gives the maximum duration of high sea levels and Table LV the maximum duration of low sea levels.

The figures in Table LII and LIII show the average number of consecutive records for every fourth hour where a given height of sea level has been reached or passed. By multiplying these figures by 4 we obtain the approximate mean duration of the different heights in hours. A comparison of the two tables indicates that although the low heights

TABLE L

THE AVERAGE NUMBER OF HOURS PER YEAR DURING WHICH A HIGH SEA LEVEL HAS BEEN REACHED OR PASSED

Sea-level height (cm):	50	55	60	65	70	75	80	85	90	95	100	105	110	115	120
Average no. of hours per year	125.9	82.7	52.5	32.0	18.4	10.7	7.5	4.3	2.7	1.6	0.6	0.5	0.3	0.2	0.2
Dec.–Jan.	67.4	43.8	28.3	15.9	9.4	5.2	3.5	2.2	1.5	0.9	0.3	0.2	–	–	–
%	53.5	53.0	53.9	49.7	51.1	48.6	46.7	51.2	55.6	56.2	50.0	40.0	–	–	–

TABLE LI

THE AVERAGE NUMBER OF HOURS PER YEAR DURING WHICH A LOW SEA LEVEL HAS BEEN REACHED OR PASSED

Sea-level height (cm):	−50	−55	−60	−65	−70	−75	−80	−85	−90
Average no. of hours per year	72.4	42.3	22.1	9.9	4.6	2.4	1.6	0.9	0.2
Feb.–April	46.9	26.1	12.8	5.0	2.2	1.2	1.0	0.4	0.1
%	64.8	61.7	57.9	50.5	47.8	50.0	62.5	44.4	50.0

TABLE LII

THE AVERAGE DURATION OF HIGH SEA LEVELS AT HELSINKI, PERIOD 1904–1953*

Sea-level height (cm):	50	55	60	65	70	75	80	85	90	95	100	105	110	115	120
Jan.	3.4	3.0	2.7	2.4	2.2	1.7	1.6	1.5	1.5	2.0	–	–	–	–	–
Febr.	2.6	2.3	1.9	1.6	1.8	1.6	1.3	1.0	–	–	–	–	–	–	–
March	2.3	2.2	1.7	2.2	1.6	2.0	2.0	1.7	1.5	1.0	1.0	1.0	1.0	–	–
April	3.5	2.7	2.2	2.0	2.0	1.0	1.0	–	–	–	–	–	–	–	–
May	2.0	2.0	1.0	–	–	–	–	–	–	–	–	–	–	–	–
June	–	–	–	–	–	–	–	–	–	–	–	–	–	–	–
July	2.2	2.5	2.0	2.0	1.0	1.0	1.0	–	–	–	–	–	–	–	–
Aug.	2.9	1.9	2.8	2.7	2.0	1.5	–	–	–	–	–	–	–	–	–
Sept.	2.7	2.2	1.8	1.8	1.6	2.0	1.3	2.0	2.0	1.0	–	–	–	–	–
Oct.	3.4	2.4	1.9	1.6	1.3	1.4	1.4	1.0	1.0	–	–	–	–	–	–
Nov.	4.1	3.7	3.4	3.1	2.4	2.8	2.4	1.8	1.6	1.2	1.5	2.0	2.0	2.0	2.0
Dec.	4.6	3.5	3.0	2.9	2.4	3.1	3.3	3.0	2.6	1.6	1.3	1.5	–	–	–
Year	3.1	2.6	2.2	2.2	1.8	1.8	1.7	1.7	1.7	1.4	1.3	1.5	1.5	2.0	2.0

* The figures give the average number of consecutive records for every fourth hour where a given height has been reached or passed. By multiplying these figures by 4 the average duration in hours is obtained. This also applies to Table LIII–LV.

TABLE LIII

THE AVERAGE DURATION OF LOW SEA LEVELS AT HELSINKI, PERIOD 1904–1953*

Sea-level height (cm):	−50	−55	−60	−65	−70	−75	−80	−85	−90
Jan.	4.7	3.6	2.7	2.6	2.2	2.0	2.0	2.0	1.0
Feb.	6.0	4.3	3.9	2.2	3.0	4.0	3.0	1.0	−
March	6.4	7.0	4.0	2.7	4.2	3.0	2.4	3.0	1.0
April	5.1	4.4	3.3	2.5	4.0	3.0	3.0	1.0	−
May	5.2	3.4	2.0	2.0	−	−	−	−	−
June	1.0	−	−	−	−	−	−	−	−
July	−	−	−	−	−	−	−	−	−
Aug.	2.0	−	−	−	−	−	−	−	−
Sept.	2.0	1.0	−	−	−	−	−	−	−
Oct.	2.6	4.4	2.5	4.0	4.0	3.0	2.0	2.0	−
Nov.	3.8	3.3	2.8	2.5	1.5	1.0	−	−	−
Dec.	6.0	6.7	6.0	4.6	4.3	3.5	3.0	1.0	−
Year	4.1	4.2	3.4	2.9	3.3	2.8	2.5	1.7	1.0

*See note below Table LII.

were less frequent than the corresponding high values, their average duration was in general considerably more prolonged. There is, of course, a decreasing tendency in the duration values in Table LII and LIII towards the more extreme values of the departures. Nevertheless, the general course is by no means smooth or uniform; a considerable number of exceptions to the main tendency may be noted. They are due to the fact that with increasing, and decreasing heights the frequencies diminish rapidly and a few fairly exceptional cases of comparatively long duration may become more and more predominant.

The maximum duration of a definite sea-level height is a factor which may be of considerable interest in different respects, above all technical and practical ones. In com-

TABLE LIV

THE MAXIMUM DURATION OF HIGH SEA LEVELS AT HELSINKI, 1904–1953*

Sea-level height (cm):	50	55	60	65	70	75	80	85	90	95	100	105	110	115	120
Jan.	20	14	14	6	5	3	3	3	2	2	−	−	−	−	−
Feb.	9	5	4	4	2	2	2	1	−	−	−	−	−	−	−
March	5	4	3	3	3	3	3	3	2	1	1	1	1	−	−
April	5	4	3	2	2	1	1	−	−	−	−	−	−	−	−
May	2	2	1	−	−	−	−	−	−	−	−	−	−	−	−
June	−	−	−	−	−	−	−	−	−	−	−	−	−	−	−
July	4	4	2	2	1	1	1	−	−	−	−	−	−	−	−
Aug.	8	5	5	4	3	2	−	−	−	−	−	−	−	−	−
Sept.	8	6	5	3	3	3	2	2	2	1	−	−	−	−	−
Oct.	20	11	10	5	2	2	2	1	1	−	−	−	−	−	−
Nov.	52	30	22	20	9	9	7	3	3	2	2	2	2	2	2
Dec.	35	24	13	10	10	9	9	8	4	3	2	1	−	−	−

*See note below Table LII.

TABLE LV

THE MAXIMUM DURATION OF LOW SEA LEVELS AT HELSINKI, 1904–1953*

Sea-level height (cm):	−50	−55	−60	−65	−70	−75	−80	−85	−90
Jan.	15	13	8	6	3	2	2	2	1
Feb.	29	18	17	7	5	4	3	1	−
March	54	25	17	7	6	5	4	3	1
April	37	30	12	5	4	3	3	1	−
May	17	9	3	2	−	−	−	−	−
June	1	−	−	−	−	−	−	−	−
July	−	−	−	−	−	−	−	−	−
Aug.	2	−	−	−	−	−	−	−	−
Sept.	2	1	−	−	−	−	−	−	−
Oct.	21	10	7	4	4	3	2	2	−
Nov.	11	11	6	4	2	1	−	−	−
Dec.	40	35	15	8	7	6	3	1	−

*See note below Table LII.

mon with Table LII and LIII, all figures in Table LIV and LV must be multiplied by 4 in order to obtain approximately the maximum duration in hours.

Table LIV shows considerable dispersion in the figures. For instance, there was in November a period covering 208 hours with a sea level at least 50 cm above the mean sea level and generally higher, while in October the corresponding duration was only 80 hours, and in December 140 hours. This shows that really long continuous periods with more-or-less extreme sea-level heights are rare. It is, moreover, characteristic of these long periods with extreme sea-level departures that they do not show a general build-up to a single peak. On the contrary, it appears quite distinctly from the groups of height 55 and 60 cm or more for November that there were two marked peaks during the above-mentioned long period of 208 hours. The same also concerns the other periods. Similar remarks are applicable to Table LV.

Although the data in Table LII–LV refer to Helsinki, they can also be extended to cover other sea-level stations along the Finnish coast. A more detailed examination has shown that there exists a fairly pronounced relationship between the frequencies of the different sea-level heights at the particular stations, and that it is the position of these stations along the coast that is the decisive factor. Minor corrections may in some cases be necessary, but they are left out of consideration in the following discussion. In order to determine the approximate frequencies in terms of the average number of hours per year for the different stations in Finland, the numerical results for Helsinki must be multiplied by a given factor. The approximate value of this factor, which is the same for high and low sea levels, is given in Table LVI.

This table reveals, for instance, that the frequencies may be considered identical at Helsinki and at Vaasa if statistical information is required. A comparison with the more elaborate Table XLVIII indicates the accuracy of the ratios in Table LVI.

The frequencies and the duration of different sea-level heights are, as already mention-

TABLE LVI

FACTORS GIVING, FOR DIFFERENT HEIGHT DEPARTURES FROM THE AVERAGE SEA LEVEL, THE RELATIONSHIP BETWEEN THE FREQUENCIES AT PARTICULAR SEA-LEVEL STATIONS ALONG THE FINNISH COAST AND HELSINKI

Height departures (cm)	Kemi	Oulu	Raahe	Pietarsaari	Vaasa	Kaskinen	Mäntyluoto	Rauma	Degerby	Turku	Hanko (Hangö)	Helsinki	Hamina
50	3.5	2.5	2.0	1.5	1.0	0.8	0.6	0.5	0.4	0.6	0.6	1.0	1.5
60	5.0	4.0	3.0	2.0	1.0	0.7	0.5	0.4	0.2	0.4	0.5	1.0	2.5
70	7.0	5.0	4.0	2.0	1.0	0.7	0.5	–	–	0.4	0.4	1.0	3.0
80	13.0	9.0	6.0	2.5	1.0	–	–	–	–	–	–	1.0	5.0

TABLE LVII

YEARLY FREQUENCIES OF CASES WITH AN HOURLY SEA-LEVEL DIFFERENCE GREATER THAN 5, 10, 15, 20, 25 AND 30 CM AT KEMI, DEGERBY AND HAMINA

	> 5 cm	> 10 cm	> 15 cm	> 20 cm	> 25 cm	> 30 cm
Kemi	295.5	34.0	8.4	2.9	1.2	0.4
Degerby	67.9	4.1	0.5	–	–	–
Hamina	310.8	24.3	2.4	0.5	–	–

ed above, important terms for many technical purposes. Nevertheless, they are not sufficient in all cases. Sometimes the speed with which the changes in sea level may occur is
the factor which is required. As with the frequencies of sea-level heights, the maximum
speed with which variations in sea level may occur is of greatest interest. In this connection there arises the question of the time interval which should be selected for the
determination of the rate of sea-level variations. Theoretically a shorter interval is doubtless preferable, but for practical purposes a more prolonged period may offer considerable
advantages. It is obvious that for harbour and other coastal constructions it is more
important to know the absolute heights of sea-level changes than the rate of these variations, let us say, during an hour. On the other hand, it must be borne in mind that rapid
sea-level changes usually cover a period of one-half to one day and more, although the
most pronounced cases of increase and decrease, forming only a part of a more prolonged
process, are of short duration. This duration, which varies within wide limits, is difficult
to determine statistically. For study of the maximum speed of sea-level changes an
interval of one hour seemed therefore to be the most appropriate.

The study of the occurrence of different speeds of sea-level variation in the Baltic Sea
along the Finnish coast (Lisitzin, 1952) was based on the three stations Kemi, Degerby
and Hamina which represent the extreme maximum (Kemi and Hamina) and minimum
(Degerby) changes. Table LVII gives the results obtained as the number of cases with an
hourly increase or decrease in sea level exceeding 5, 10, 15, 20, 25 and 30 cm.

A comparison of the data for the particular stations shows that a considerable difference exists between Kemi and Hamina on the one hand and Degerby on the other hand.
It may be noted that while the yearly number of hourly sea-level differences from 5 cm
upwards at the former stations is roughly 300, corresponding approx. to 3.5% of all
observations, the frequency at Degerby is 68 cases or only 0.8%. Table LVII indicates,
moreover, the inequality between Kemi and Hamina. This inequality is already perceptible for differences greater than 10 cm and it increases rapidly when progressing towards
more marked differences. The extreme values also deviate highly. The most pronounced
hourly difference is at Kemi 42 cm and at Hamina only 23 cm. For Degerby this difference is 18 cm.

It has already been pointed out that a statistical examination of the rapid sea-level
variations within an interval of one hour is not sufficient to provide a satisfactory answer
to many technical problems. In order to complete the results given above, the frequency
of days with an absolute sea-level difference of 0.1–10.0 cm, 10.1–20.0 cm, etc. was
determined. The stations used for this purpose were Kemi, Raahe, Mäntyluoto, Degerby
and Hamina and the research period was 1940–1949. The results obtained, expressed in
percentages, are reproduced in Table LVIII.

The table shows the varying character of the daily sea-level changes along the Finnish
coast. At Degerby almost two-thirds of these variations occur within the limits of 10 cm,
and roughly 94% reach at the utmost the height difference of 20 cm. In addition, it may
be mentioned that at this station the daily differences are without exception below
50 cm. At Mäntyluoto a distinct displacement of the daily sea-level differences towards

TABLE LVIII

THE AVERAGE FREQUENCIES (IN PERCENTAGES) OF DIFFERENT DAILY SEA-LEVEL DIFFERENCES

Sea-level differences (cm):	0.1–10.0	10.1–20.0	20.1–30.0	30.1–40.0	40.1–50.0	50.1–60.0	60.1–70.0	70.1–80.0	80.1–90.0	90.1–100.0	100.1–110.0	110.1–120.0	120.1–130.0	130.1–140.0	140.1–150.0
Station:															
Kemi	25.0	36.0	18.7	9.2	4.8	2.8	1.8	0.7	0.7	0.14	0.14	0.05	0.09	–	–
Raahe	28.8	39.3	16.8	8.1	3.5	2.0	0.9	0.2	0.2	0.07	0.03	–	–	–	–
Mäntyluoto	52.5	34.0	9.3	2.7	1.0	0.2	0.1	0.03	–	–	–	–	–	–	–
Degerby	63.4	30.1	4.8	1.2	0.3	–	–	–	–	–	–	–	–	–	–
Hamina	22.0	40.7	18.0	8.8	5.1	2.4	1.2	0.9	0.5	0.2	0.2	0.07	0.03	0.07	0.03

TABLE LIX

THE SEASONAL CYCLE OF THE AVERAGE DAILY SEA-LEVEL DIFFERENCES (CM)

Station	J	F	M	A	M	J	J	A	S	O	N	D	Year
Kemi	28.6	17.0	15.7	15.0	17.3	16.5	13.0	17.7	23.4	29.6	28.6	29.1	21.0
Raahe	22.9	17.6	15.8	14.5	14.3	13.0	11.1	15.3	17.8	24.2	25.5	25.5	18.1
Mäntyluoto	16.1	13.1	12.6	10.9	9.2	8.5	7.1	9.6	11.6	14.8	15.2	16.2	12.1
Degerby	12.7	11.1	11.1	9.4	8.5	8.1	6.8	8.0	9.3	11.2	12.0	12.7	10.1
Hamina	28.6	22.1	21.5	17.2	16.2	15.4	12.7	17.2	18.2	23.3	29.0	28.0	20.8

TABLE LX

THE SEASONAL CYCLE OF THE MAXIMUM DAILY SEA-LEVEL DIFFERENCES (CM)

Station	J	F	M	A	M	J	J	A	S	O	N	D	Year
Kemi	103	64	63	65	101	59	86	88	88	123	100	112	123
Raahe	86	70	86	82	102	50	62	69	73	96	71	94	102
Mäntyluoto	64	59	70	52	39	32	24	32	38	62	71	56	71
Degerby	45	48	38	32	28	25	18	25	34	48	47	41	48
Hamina	109	86	84	79	70	61	59	107	83	103	133	142	142

higher values may be noted. This appears not only from the decreasing percentage of cases with differences below 10 cm, but also from the increasing value for the other groups and the occurrence of one case with the daily sea-level difference exceeding 70 cm. A similar trend, still more pronounced, is characteristic of Raahe and Kemi. When considering Hamina in the Gulf of Finland it may be noted that this station shows in many respects an amazing similarity to Kemi. The most marked features of the values for Hamina are that the percentage of days with small differences is at this station at its lowest and the maximum of the daily differences at its highest, reaching in one case as high a value as 142 cm.

The principal aim of Table LIX is to give a concept of the average seasonal cycle of daily sea-level differences. This table comprises the average daily sea-level differences during particular months for the same stations and periods as in Table LVIII, while the corresponding recorded maximum differences are given in Table LX.

Table LIX shows distinctly the pronounced deviations between the individual months. At all stations the minimum height difference is reached in July. At Kemi there is a secondary minimum in April which must probably be ascribed to the influence of the ice cover. The time for the occurrence of the maximum is not as regular and varies over a period covering 4 months, October to January. However, it is not impossible that the early maximum for Kemi is also a consequence of the shift of extreme sea-level differences due to the considerable extent of the ice cover during the winter months. Finally, it may be pointed out that the ratio between the maximum and the minimum for the same station shows a remarkable regularity. Thus for all the stations, with the exception of Degerby, the minimum amounts to approximately 44% of the maximum; for the latter station it is roughly 54%.

Examining Table LX more closely, it may also be noted that in connection with the maximum sea-level differences the summer months are characterized by the lowest values, while the more marked differences in sea-level occur generally in autumn and winter. The secondary maximum at Kemi and Raahe in May forms an exception to this rule and indicates the highly irregular character of the phenomenon and the difficulty of drawing general conclusions.

For consideration of the more prolonged periods, it is reasonable to select one year as the time unit. The maximum, average and minimum values of sea-level differences during one year are reproduced in Table LXI. The fluctuations of the deviations are once more most accentuated in the northern parts of the Gulf of Bothnia, where at Raahe the maximum departure is almost 2.2 times larger than the minimum one. Southwards the ratio diminishes, amounting for instance to 1.6 only at Helsinki.

Finally, a comparison of the ratio between the corresponding amplitudes for the particular stations may be of considerable interest. As immediately comparable figures are indispensable in this connection, the sea-level differences for the separate series at Degerby were denoted by 1.00 and the relative differences of the other stations determined starting from this value. Table LXII shows these relative values for the average daily, monthly and yearly differences as well as for the absolute amplitudes of sea-level depar-

TABLE LXI

THE MAXIMUM, AVERAGE AND MINIMUM VALUES OF THE YEARLY SEA-LEVEL DIFFERENCES (CM)

	Kemi	Oulu	Raahe	Pietar-saari	Vaasa	Kaski-nen	Mänty-luoto	Rauma	Turku	Deger-by	Hanko (Hangö)	Helsin-ki	Hamina
Maximum	282	258	252	215	192	179	156	142	156	131	146	184	243
Average	214	192	182	153	139	129	117	112	116	96	114	145	184
Minimum	144	123	116	99	89	90	84	85	77	64	89	112	144

TABLE LXII

THE RATIOS OF MEAN SEA-LEVEL DIFFERENCES FOR DIFFERENT PERIODS BETWEEN PARTICULAR STATIONS AND DEGERBY

	Kemi	Oulu	Raahe	Pietar-saari	Vaasa	Kaski-nen	Mänty-luoto	Rauma	Turku	Deger-by	Hanko (Hangö)	Helsin-ki	Hamina
Mean daily differences	2.08	–	1.79	–	–	–	1.20	–	–	1.00	–	–	2.06
Mean monthly differences	2.31	2.11	1.93	1.62	1.47	1.33	1.24	1.22	1.24	1.00	1.20	1.56	1.98
Mean yearly differences	2.33	2.00	1.90	1.59	1.45	1.34	1.22	1.17	1.21	1.00	1.19	1.51	1.92
Total differences	1.99	1.93	1.86	1.61	1.36	1.25	1.16	1.09	1.19	1.00	1.16	1.32	1.71

tures. With the exception of the daily deviations the amplitudes show a relative decrease when passing from shorter periods to longer ones. The correlation is at its lowest for the stations situated in the innermost parts of the gulfs, where the distance to Degerby is greatest. The relative deviations corresponding to different periods there reach values which differ by about 15%. These departures are, nevertheless, not marked enough to disturb the general character of the phenomenon as a whole.

Since statistics on sea-level variations and their characteristics are frequently compiled by different technical bodies and offices and distributed by these authorities on a professional basis, they are usually not easily accessible. The results obtained for the Finnish coastal stations have therefore been largely utilized in order to illustrate the principal features of the changes concerned. The separate tables may also give an idea of the possibilities and procedures needed in treating sea-level records statistically for use in planning technical constructions and other similar projects in coastal regions. Corresponding statistics have assuredly been compiled, or at least may be compiled, for all coastal areas.

Some examples of sea-level statistics based on average values for a definite span of time have already been given. The monthly sea-level values for Brest are very extensive, since the observations started there as early as in 1807. It may therefore be interesting to examine statistically the frequencies of the monthly sea-level heights for a prolonged period, i.e., the 123-year period covering the years 1807–1835, 1846–1856 and 1861–1943 (Lisitzin, 1955b). The average monthly sea-level heights (in cm) were determined for Brest for the years enumerated above, relative to the mean sea level for the whole period, and were as follows:

J	F	M	A	M	J	J	A	S	O	N	D
0.8	−2.0	−3.5	−3.2	−2.7	−3.7	−3.1	−1.7	0.7	7.2	6.7	4.3

These data show that the largest departures between the monthly averages is 10.9 cm, while the variations during the period from March to July do not exceed 1 cm.

The picture changes considerably if monthly means for particular years are studied in more detail. In these cases the entire phenomenon, which is probably characteristic of a relatively extensive area, changes as a consequence of different regional meteorological factors. The departures are fairly pronounced during separate years. The highest monthly sea level, relative to the mean sea level of the corresponding year, was noted for December 1876, reaching a height of 34.9 cm, while the lowest monthly sea level was observed in May 1912, showing a negative deviation of 23.5 cm. In this connection it must, however, be pointed out that the maximum mentioned above represents a rather exceptional case, being 5.6 cm higher than the next highest value. In Table LXIII are given the cases where the monthly sea level at Brest deviated by at least 20 cm from the mean sea level of the corresponding year.

Table LXIII shows that all the high sea levels were observed in autumn or in winter, i.e., during the period from October to February, while the low sea levels, which are considerably less frequent, occurred during the months February to May. Concerning the

TABLE LXIII
THE MONTHLY SEA-LEVEL HEIGHTS AT BREST DEVIATING BY MORE THAN 20 CM FROM
THE MEAN SEA LEVEL OF THE CORRESPONDING YEAR

Time	High sea levels (cm)	Time	Low sea levels (cm)
Dec. 1876	34.9	May 1912	−23.5
Dec. 1868	29.3	Feb. 1934	−23.3
Dec. 1821	28.1	Feb. 1852	−23.2
Dec. 1934	25.8	April 1912	−21.4
Nov. 1852	25.0	Feb. 1891	−21.4
Feb. 1912	24.8	March 1813	−21.2
Oct. 1822	24.2	April 1817	−20.3
Dec. 1914	23.6		
Jan. 1936	23.6		
Jan. 1873	22.8		
Feb. 1936	22.4		
Oct. 1813	21.3		
Dec. 1910	20.8		
Nov. 1926	20.6		
Oct. 1824	20.4		

regional coverage of the extreme monthly sea-level heights, it could be established that a high or low monthly sea level in southwestern England could generally be noted simultaneously with Brest, but that the correlation was, as a rule, weak on the Dutch coast.

In order to obtain a concept of the seasonal distribution of different monthly sea-level heights, the levels corresponding successively to the frequencies 5, 10, 15, etc. up to 90 and 95% were determined. The results are reproduced in Table LXIV, completed by the extreme values for every month. Some explanations may be necessary. For instance, the frequency of 5% for July indicates that the sea level 3.8 cm above the yearly level was reached or exceeded in 5 years during a century. In December the given sea level was reached or exceeded in half of all cases, the frequency being 50%. Finally, it may be noted that in October the frequency of the named sea level is still more pronounced, 68%. On the other hand, the frequency corresponding to −3.5 cm is 95% in November, but only 47% in June. Concerning the extreme heights it may be pointed out that in July the highest observed monthly sea level is only 5.7 cm, being thus approximately 6 times less than in December, when it was 34.9 cm. For the minimum sea levels the differences in departures for the particular months are less marked. However, it may be noted that the lowest monthly sea level observed in October is 8.9 cm below the yearly sea level, while in May the value is numerically 2.6 times higher, reaching −23.5 cm.

Moreover, attention should be paid to the possible largest monthly range of sea level variations at Brest. This range is determined by the following series:

J   F   M   A   M   J   J   A   S   O   N   D
41.0 48.1 37.1 35.7 34.5 23.2 19.7 21.0 27.9 33.1 36.8 51.6

In spite of a great number of deviations the pronounced difference between the winter

TABLE LXIV

THE FREQUENCY OF MONTHLY SEA-LEVEL HEIGHTS (CM) AT BREST DURING DIFFERENT MONTHS

%	J	F	M	A	M	J	J	A	S	O	N	D
Max.	23.6	24.8	16.0	14.3	11.0	8.8	5.7	9.1	17.3	24.2	25.0	34.9
5	16.2	14.9	9.1	9.5	7.1	4.3	3.8	5.7	11.2	18.6	17.6	20.6
10	12.6	11.9	5.9	5.9	4.5	2.2	1.6	4.1	7.4	16.2	15.5	16.9
15	9.9	9.4	4.4	3.7	3.3	0.1	0.7	2.5	6.5	14.9	13.6	13.5
20	8.2	6.6	2.5	1.4	1.4	-0.9	0.2	1.6	4.8	13.2	12.5	12.2
25	6.6	3.6	1.8	0.7	0.7	-1.3	-0.6	1.2	3.6	12.0	10.7	10.0
30	5.1	2.5	0.0	-0.6	-0.2	-1.8	-1.2	0.0	2.8	11.0	10.1	7.9
35	4.2	0.6	-1.0	-1.5	-0.5	-2.3	-1.7	-0.4	2.0	10.1	9.4	7.4
40	3.5	0.0	-1.5	-1.8	-1.1	-2.7	-2.1	-1.2	1.7	9.0	9.1	6.5
45	1.9	-1.4	-2.7	-2.8	-2.0	-3.3	-2.7	-1.7	1.0	7.8	8.0	5.3
50	0.4	-2.2	-3.6	-3.6	-2.4	-3.8	-3.1	-2.2	0.2	7.0	7.3	3.8
55	-0.3	-4.3	-4.4	-4.7	-3.2	-4.3	-3.4	-2.5	-0.1	6.0	6.5	2.9
60	-1.3	-5.0	-5.0	-5.2	-3.6	-4.5	-3.9	-2.9	-0.8	5.3	5.7	1.9
65	-2.5	-6.0	-5.5	-5.6	-4.7	-5.1	-4.4	-3.4	-1.5	4.5	4.7	-0.1
70	-3.5	-8.1	-6.2	-6.1	-5.6	-5.8	-4.9	-3.8	-2.6	3.6	3.3	-0.8
75	-5.7	-9.0	-7.6	-6.5	-6.1	-6.2	-5.5	-4.3	-2.9	2.6	2.4	-1.7
80	-8.4	-9.3	-9.8	-7.5	-6.6	-6.9	-6.2	-5.1	-4.0	0.9	1.0	-2.8
85	-9.0	-10.9	-11.1	-8.4	-7.5	-8.3	-7.0	-5.7	-4.5	0.2	-0.5	-5.3
90	-10.3	-14.0	-11.9	-10.2	-9.3	-9.4	-8.1	-6.2	-5.0	-1.7	-2.3	-7.5
95	-12.8	-15.1	-13.3	-12.9	-11.0	-11.0	-9.5	-8.0	-6.6	-4.0	-3.5	-10.2
Min.	-17.4	-23.3	-21.1	-21.4	-23.5	-14.4	-14.0	-11.9	-10.6	-8.9	-11.8	-16.7

months (December to February) and the summer season (June to August) may be noted, while spring and autumn represent the transitory periods. These results prove the contribution of meteorological effects, especially of the wind, on the seasonal distribution of the monthly sea-level heights.

STORM-SURGE FORECASTS

After the disastrous floods of January and February 1953 in England, The Netherlands and Belgium a Flood Warning Service came into operation in Great Britain and has been operating since then during each winter from about the middle of September to the end of April.

The Service bases its work partly on the weather forecast received from the Meteorological Office and partly on reports of sea-level heights recorded at Stornoway, Aberdeen, Tyne, Immingham, Lowestoft and Harwich, completed with predictions of the hourly heights of the astronomical tide prepared in advance for these ports. The recorded heights of sea level are received by the Service once daily even during periods of calm weather, mainly in order to check that the communication system is working properly. During periods when storm surges may be expected, the sea-level data are received once every hour (Proudman, 1963).

The method of prediction is empirical. It is related to the values of atmospheric pressure or wind at fixed points at sea for the same or earlier occasions. The relationship between sea level and the meteorological conditions during past storm surges is derived by statistical analysis. For the determination of the storm-surge height at a given location the heights of the disturbance at other locations struck at an earlier time point are taken into consideration. Corkan (1948), on the basis of an analysis of a number of selected storm surges, developed for the east coast of Great Britain an empirical formula, which is reproduced in the following — transformed from feet into centimetres — in order to give an example of the approach to the problem. The formula of Corkan reads:

$$R_S = R_D + 1.01\, N\,|N| - 1.68\, E\,|E| - 2.29\, n\,|n| - 2.90\, e\,|e|$$

where R_S is the disturbance in sea level at Southend after the elimination of the local atmospheric pressure assuming a statical law, at the time t hours, and R_D is the recorded disturbance at Dunbar after the elimination of the effect of the local atmospheric pressure assuming in this case also a static law, at the time $t - 9$ hours. N and E stand for the north and east pressure gradients at a point A in the southern part of the North Sea — situated approximately halfway between Southend and Hook of Holland, but some distance to the north — at time t hours. n and e are the north and east pressure gradients at a point B close to the centre of the North Sea between Aberdeen and Esbjerg, at the time $t - 6$ hours. The atmospheric pressure gradient at a point is measured as the difference in pressure in mbar over a distance of 250 miles centered on the point.

The Corkan formula allows a storm-surge forecast to be made for Southend not more than 9 hours beforehand. The terms involving N and E represent the storm surge at Southend produced by local winds, while the terms involving n and e represent the storm surge at Southend as a consequence of winds acting over the whole extension of the North Sea.

For practical reasons Aberdeen was later substituted for Dunbar. On average it may be estimated that it takes a storm surge roughly two hours to proceed from Aberdeen to the Tyne, another two hours to reach Immingham, an additional four or five hours to Lowestoft and another two or three hours before Southend is reached. Observations on sea level are made at all these places, so that all estimates may be checked and corrected as the storm surge progresses on its way to the south.

Empirical methods for the forecasting of storm surges have also been developed for the German North Sea coast and applied as a matter of routine work by the Deutsches Hydrographische Institut. Similar procedures have also been used for deriving formulae for the prediction of storm surges at different localities along the Atlantic Ocean and Gulf coasts of the United States and along the Japanese coast.

In addition to the purely empirical approaches to the problem of storm-surge forecasting, there are also theoretical methods based on the integration of the hydrodynamic differential equations which represent the generation and development of storm surges. As far as is known, these methods have not so far been utilized for the practical forecasting of surges on a large scale. Although an electronic computer may be of great help in this connection in order to determine the response of the sea to the effects of the meteorological factors, there are at the present time considerable drawbacks in the method arising, for instance, from the uncertainties in the determination of the surface wind stress from available meteorological data, and lack of proper knowledge of the values of the bottom friction coefficient. Nevertheless, these methods, if given good enough computing facilities, possess a great potential for dealing with the complicated variations in atmospheric pressure and wind fields which occur in actual cases in nature, and the possibility of investigating phenomena produced by these variations at sea.

The advantages of the empirical methods on the one hand and the theoretical approaches to the problem on the other hand have been discussed in several connections. It must always be borne in mind that the empirical method is based upon a statistical analysis of the relationship between the meteorological elements and the reaction of the sea level to the concerned effects. Since less-pronounced storm surges are by far the most frequent, there will always be a tendency to give a greater weight to these cases in the statistical computations, and the formulae derived in this way may be less adequate in connection with marked storm surges, which, of course, are the most significant in practice. The use of an empirical formula is, moreover, based on a restricted amount of data for atmospheric pressure and wind and therefore will hardly be fully representative in all cases. The weakness of the more theoretical procedures, principal purpose of which is to achieve sufficient accuracy of predicted sea-level heights caused by storm surges, is the difficulty of translating the physical character of the meteorological phenomenon and the properties of sea water into mathematical forms appropriate for the calculations.

THE TSUNAMI WARNING SYSTEM

For a long time — according to some sources since 1896 — a system providing tsunami warning to the inhabitants has existed in Japan. The earthquake followed by tsunami waves which occurred on April 1, 1946, and resulted in a disastrous loss of life and property in the Hawaiian Islands emphasized the necessity of a similar system in the United States and forced the authorities to take measures for its establishment. The Coast and Geodetic Survey of the United States Department of Commerce accepted responsibility for supervising and operating this warning system in the Pacific Ocean area. Not only military services and other governmental bodies, but also private organizations were willing to offer their collaboration. As the result of a careful examination a convenient network of strategically situated seismic and sea-level stations was set up as the nucleus for the Pacific warning system. The cooperation of several countries was easily secured for participation in this work. All necessary information from the separate stations in the warning net are communicated to the system's headquarters in Honolulu.

Originally, the principial purpose of the warning system was to provide tsunami information for the population in Hawaii. However, at a relatively early stage the system was also extended to cover other United States military bases in the whole area of the Pacific Ocean. A few years later many foreign countries and the major islands in the Pacific Ocean were incorporated in the system. The disastrous tsunamis which followed the earthquake in Kamchatka on November 4, 1954, and the Chilean earthquake on May 22, 1960, have very distinctly shown the efficiency of the warning system. The warnings reached not only the authorities but also the population sufficiently in advance, i.e., before the arrival of the tsunami waves, that a number of protective measures could be taken which limited the inevitable loss of life and property to a minimum. Although the results have been mainly positive, there is still a definite necessity to improve and develop the warning system in many respects. The governments in the particular countries affected should try to intensify their efforts in educating the local population in the Pacific area regarding the danger connected with the occurrence of tsunamis.

One of the serious intrinsic limitations of an effective warning system on a large scale is the fact that it is not able to provide efficient warning to countries or islands situated in the vicinity of the epicentres of the submarine earthquakes. This danger is especially pronounced in Japan. It was therefore of the greatest significance that participants should try to develop their own warning systems and a service capable of detecting the tsunami source active in the vicinity of the coasts, in order to warn the population without delay. If such a service is not maintained and the population is dependent upon the warning information from Honolulu, a sense of security which is not real may easily arise. Japan and the Soviet Union therefore have their own centres, where in addition to the tsunami warning activities intensive scientific studies of the phenomenon are carried out. This part of the work was initiated in the Soviet Union in 1952 (Soloviev, 1972) and in Japan in 1960 (Uda, 1971).

A few words should also be dedicated in this connection to the practical background and operation of the warning system. The main activities of the Honolulu Observatory are to supervise the operational control of the system, to detect submarine earthquakes, determine the position of their epicentres, request and obtain additional information, evaluate the reports concerning the sea waves and issue warnings to the different authorities whenever the need of such measures is evident. The contribution of the seismic stations consists in detecting the submarine earthquakes and in sending necessary reports to the Honolulu Observatory. Some of the seismic stations also submit the seismographic reports concerned to Tokyo. The sea-level stations, which are considerably more numerous than the seismic stations, have to detect sea waves by means of recording gauges or automatic sea-wave recorders and to send their reports to the Honolulu Observatory on request or, if necessary, on their own initiative.

As soon as the seismographic report has reached Honolulu Observatory, it is examined without delay in order to determine the position of the epicentre of the earthquake. If the epicentre is situated in an area in the Pacific Ocean favourable for the generation of a tsunami wave, the Observatory sends messages to sea-level stations which the wave may reach in the course of its progress. These stations have to watch the records of their gauges very carefully for evidence of the arrival of the tsunami wave and to report the findings immediately to the Honolulu Observatory.

Sometimes it may occur that the arrival of the tsunami wave is the first sign of the existence and propagation of such a wave. For this purpose the Coast and Geodetic Survey has developed a 'seismic sea-wave detector'. This apparatus automatically screens out wind-caused waves and astronomical tides. It responds only to oscillations with the period of seismic sea waves — 10–40 min — and by closing an electric circuit alerts the observer, who examines the record and reports the results of this examination to Honolulu.

All wave reports received by the Honolulu Observatory are immediately investigated by the staff. If the analysis of the data received indicates that further action is imperative, the Observatory will promptly send warning information to all authorities concerned urging them to every action necessitated by the circumstances and permitted by the time before the expected arrival of the tsunami wave.

Honolulu Observatory has the complete right to cancel all advisory and warning actions by notifying the previously designated authorities.

In this connection the complex character of the tsunami waves may be pointed out once more. Not all submarine earthquakes produce sea waves and not all tsunami waves cause damage. However, it must always be kept in mind that although a wave may not appear to be dangerous at one station, it may exert a disastrous effect at another station owing to the influence produced by the shape of the coastline or the configuration of the ocean floor in the vicinity of the shore. It is therefore of the greatest importance that all suspected waves are closely examined by the Honolulu Observatory.

SEA-LEVEL CHANGES AND WATER POLLUTION

The danger of water pollution in oceans and seas and in particular in coastal regions is continuously increasing, and the inevitable consequence is the growing significance of investigating such questions as water mixing, water interchange and water renewal. The problem of water pollution must thus be added to the growing list of navigational, coastal defence and engineering problems which require a more detailed understanding of water movement of every type and scale. There are at the disposal of the student several highly deviating approaches to the close investigation of this subject. Some of them require special instrumentation, extensive field work and time-consuming computations; in other cases interesting and valuable results may be reached by using data already available, since they were collected primarily for other scientific and practical purposes. For instance, data on sea-level variations recorded at tide gauges along the coasts of more-or-less enclosed bays and gulfs and in marginal seas connected with the oceans by narrow transition regions allow us to draw approximate but valuable conclusions on water interchange with the adjacent seas and oceans, and consequently also on water renewal.

In the areas where the range of the tidal variations is pronounced, the water masses are continuously renewed and the danger of pollution is therefore less marked than in regions where tides are weak, reaching generally only a few centimetres. There is no doubt that the tidal phenomenon, mainly owing to its pronounced regularity, has a certain share in the water-renewing processes. In the Bristol Channel with its marked tide, the phenomenon assuredly contributes to the purification of the water. Conversely, in comparatively shallow bays, gulfs and seas with weak tides the effect of the meteorological factors, especially the wind-produced piling-up of the water surface, is doubtless of more significance. The Baltic Sea and its extensive gulfs is a region where the interchange and renewal of the water masses may be successfully studied on the basis of already available sea-level data. Of course, more intensive research work must be done before the results can be considered conclusive. Nevertheless, the usefulness of the method on which the computations have been based is already quite obvious.

The following review is mainly based on the results obtained by Lisitzin (1967a). Some of the more important conclusions reached on the basis of the outlined approach to the problem are described in more detail.

Sea-level changes in the Baltic Sea basin are principally a consequence of the perturbating effects of different meteorological elements upon the water surface. Atmospheric pressure and wind must be mentioned as the most significant among these elements. For instance, along the coasts of Finland the part of the changes caused by tides contributes only a fraction of the total variations. A few examples may be sufficient to illustrate this fact. The height difference between the diurnal high- and low-water tides does not even in the most favourable cases exceed the following values: Kemi, 6.5 cm; Raahe, 6.5 cm; Mäntyluoto, 3.5 cm; Degerby, 3.0 cm; Hamina, 15.0 cm.

It must in addition be pointed out that the above maximum ranges occur only during fairly short periods every year in connection with a more-or-less pronounced coincidence of the extreme heights of all determined diurnal constituents. It may be added that the tide is markedly diurnal, at least in the northern and eastern parts of the Baltic.

On the other hand, a study (Lisitzin, 1952) concerning the seasonal distribution of the daily sea-level fluctuations along the Finnish coast, based on the records for the 10-year period 1940—1949, showed that the average values of the daily height differences amounted for the following stations to: Kemi, 21.0 cm; Raahe, 18.1 cm; Mäntyluoto, 12.1 cm; Degerby, 10.1 cm; Hamina 20.8 cm.

In this connection it may be mentioned that for the Gulf of Bothnia, including Degerby, these height differences are roughly six times larger than the average range of the astronomical tides, while for Hamina in the Gulf of Finland the ratio is slightly less than three. The significance of sea-level variations due to the effect of meteorological factors compared with those caused by astronomical tides is also distinctly revealed by the results for the average frequencies of daily height differences in sea level. These results show, for instance, that height deviations greater than 10 cm per day occur over a more prolonged span of years: Kemi, 75% of the days; Raahe, 71%; Mäntyluoto, 48%; Degerby, 36%; Hamina, 78%.

The continuous increase of the concerned percentage towards the inner parts of the Gulf of Bothnia and the Gulf of Finland is thus highly accentuated. The sea-level height difference limit of 10 cm, which — with the exception of Hamina — is 1.5—3.0 times greater than the maximum range of the tide, and which covers approximately 4.5—9.0 months per year, is a good indication that the contribution of the astronomical tides to the total fluctuations in sea-level heights is rather slight. Concerning Hamina it may be mentioned that daily sea-level height departures greater than 20 cm have been recorded in 59% of the days of the year and thus cover more than 7 months per annum. During the above-mentioned 10-year period, 1940—1949, the maximum differences in sea-level height during one day were for the various stations the following: Kemi, 112 cm; Raahe, 102 cm; Mäntyluoto, 71 cm; Degerby, 47 cm; Hamina, 142 cm.

These values are of a completely different magnitude than the maximum range of the diurnal tides. They therefore accentuate once more the significance of the meteorological contribution to the total fluctuations of sea level in the Baltic Sea and its extensive gulfs.

However, there is an additional point of view which must be taken into account. The extreme heights in sea level in the Baltic area are generally of short duration, covering in some cases a few hours, and in other cases only a fraction of an hour. These very short periods are hardly sufficient to allow a radical mixing of the water masses involved. It must, moreover, be borne in mind that wind-driven water circulation is restricted to the upper water layers, which, being less dense than the deeper water masses, preclude a more thorough mixing. On the other hand, bottom currents, deep-water upwelling along the coast and other compensatory processes may sometimes contribute to the acceleration of water mixing. Taking into consideration all these factors, it seemed more appropriate to choose a parameter other than the departure of sea level during one day as the starting point for the study of the renewal of the water masses in the Baltic Sea and its gulfs. The taking of the average daily sea-level height, based on six readings of the records, and its day-to-day variations must be considered as a definite improvement in the approach to the problem, since it levels out the data and thus eliminates the effect of the extreme sea-level heights of short duration. In this connection it must, however, be stressed that at the present time it is hardly possible to decide whether the day-to-day variations correspond to a period which is sufficiently short or long for the determination of the water renewal in the basin. It must always be remembered that with regard to uninterrupted sea-level fluctuations a considerable part of the original water masses may return the next day to the basin and it is therefore by no means correct to speak about a total renewal of water. The amount of water flowing to and fro is assuredly still greater if the computations for the renewal of the water masses are based on height differences in sea level for a time-span as short as, for instance, four hours. It is therefore highly inappropriate to base the determination of water renewal on such short periods. However, it must be taken into account that when we speak of the renewal of the Baltic water the deepest areas in the sea basin, such as the Gotland deep, with a water density considerably higher than elsewhere in the sea, will not participate in this renewal, unless quite specific conditions necessary for this renewal are fulfilled. Generally an extremely marked inflow of water through the transition area around Denmark, due to exceptional meteorological and hydrographic conditions, is needed to bring about such a renewal (Wyrtki, 1954). Hela (1960) roughly estimated that the average length of the renewal period for the bottom water layers in the Bornholm deep is 15 years and for the Gotland deep not less than 30 years.

The results discussed in more detail in the following text, are based on sea-level records for the 8 Finnish sea-level stations given in Fig. 46. The period of records covers 30 years from 1931–1960.

In view of the main purpose of the study, the determination of the renewal time of the water, it seemed preferable to take into account the positive day-to-day changes in sea level separately from the corresponding negative changes. It was established that the average and the maximum increase in sea level per day was generally slightly larger than the average and maximum decrease. Since, if the decrease in sea level which is the consequence of land uplift in the Baltic area is left out of consideration, increase and decrease during a prolonged period must balance each other the consequence is that there must be a slight difference in the number of days with a positive change in sea level and those characterized by a negative change. In this respect the following results were achieved:

	Increasing changes	Decreasing changes
Kemi	180 days	185 days
Degerby	173 days	192 days
Hamina	176 days	189 days

Table LXV presents the average day-to-day increase in sea level for different stations along the coasts of Finland. Considerable deviations may be noted in this table, not only in the general seasonal pattern of the increase but also in the regional distribution.

The data in Table LXV show that the average sea-level increase from one day to the

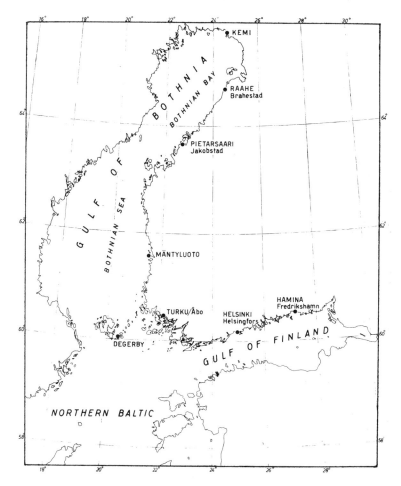

Fig. 46. The position of the Finnish sea-level stations on which the study of water pollution in the Baltic was based.

next is characterized by a fairly marked seasonal trend with a maximum late in the autumn or early winter and a minimum in July. These considerable deviations indicate that the effect of the piling-up of the water surface caused by wind is strong during the stormy season of the year. The piling-up effect may also be easily traced in the fact that the day-to-day increases in sea level are at their lowest for Degerby and increase gradually in height northward and eastward in the two large gulfs which enclose Finland. The variations are, for instance, approximately 1.5 times as large at Mäntyluoto as at Degerby, while the ratio for Raahe and Degerby is almost 3, and for Kemi and Degerby more than 3.5. The day-to-day changes are practically twice as large at Helsinki as at Degerby.

In order to give a concept of the extreme day-to-day increase in sea level during the period under review, Table LXVI has been compiled. This table confirms the above results, although it shows a less pronounced regularity in the general seasonal pattern.

TABLE LXV

THE AVERAGE DAY-TO-DAY INCREASES IN SEA LEVEL (CM)

Station	Jan.	Feb.	March	April	May	June	July	Aug.	Sept.	Oct.	Nov.	Dec.	Jan.–Dec.
Kemi	15.4	12.9	9.5	8.9	8.4	8.2	7.3	9.6	12.7	16.4	17.9	16.9	12.0
Raahe	13.9	11.0	8.7	7.9	7.0	6.4	5.6	7.1	9.7	12.1	14.3	13.9	9.8
Pietarsaari	10.7	8.7	7.6	6.5	5.6	5.2	4.7	5.5	8.1	9.8	11.2	11.0	7.9
Mäntyluoto	7.7	6.4	5.4	4.7	3.9	3.5	3.3	3.8	5.0	5.9	6.9	7.0	5.3
Turku	6.2	4.9	4.5	4.3	3.6	3.6	3.4	3.8	5.1	5.6	5.9	5.6	4.7
Degerby	4.9	3.9	3.9	3.3	2.7	2.5	2.5	2.6	3.2	3.6	4.1	4.2	3.4
Helsinki	8.3	7.5	6.4	5.9	5.3	4.8	4.5	5.4	6.9	7.4	8.2	7.4	6.5
Hamina	11.0	9.2	8.0	7.1	6.8	6.2	5.8	7.2	9.0	9.8	10.6	10.6	8.4
Average	9.8	7.7	6.8	6.1	5.4	5.0	4.6	5.6	7.5	8.8	9.9	9.6	7.2

TABLE LXVI

THE RECORDED MAXIMUM DAY-TO-DAY INCREASES IN SEA LEVEL (CM)

Station	Jan.	Feb.	March	April	May	June	July	Aug.	Sept.	Oct.	Nov.	Dec.	Jan.–Dec.
Kemi	113	79	59	70	64	52	47	47	68	83	85	72	70
Raahe	93	73	50	62	55	47	27	36	56	61	67	61	57
Pietarsaari	84	62	43	58	44	33	25	33	50	45	55	46	48
Mäntyluoto	46	36	35	34	26	20	26	28	34	43	43	32	34
Turku	37	29	22	31	20	20	25	28	46	32	41	29	30
Degerby	25	25	22	28	16	17	20	19	17	26	23	28	22
Helsinki	67	33	38	37	38	21	28	33	56	46	48	48	41
Hamina	87	54	52	40	60	32	34	42	75	60	60	67	55
Average	69	49	40	45	40	30	29	33	50	50	53	48	45

TABLE LXVII

THE SUMS OF THE AVERAGE INCREASES IN SEA LEVEL (CM)

Station	Jan.	Feb.	March	April	May	June	July	Aug.	Sept.	Oct.	Nov.	Dec.	Jan.–Dec.
Kemi	229	156	136	132	124	129	113	138	197	251	256	247	2108
Raahe	205	149	129	116	108	101	86	104	151	193	204	210	1756
Pietarsaari	161	124	112	96	85	82	74	83	123	150	161	163	1414
Mäntyluoto	113	85	80	68	58	56	51	54	75	91	101	101	933
Turku	88	65	63	60	52	55	51	53	67	80	86	85	803
Degerby	65	54	54	47	43	42	40	36	47	54	58	60	600
Helsinki	121	99	93	82	79	75	70	76	94	110	116	116	1131
Hamina	161	122	116	102	97	96	88	99	126	149	156	160	1472
Average	143	107	98	88	81	80	72	80	110	135	142	143	1279

The results mentioned above may be supplemented by the additional Table LXVII, representing the sums of the average increases in sea level during the particular months and for the whole year.

For the purpose of determining the renewal of the water in the Baltic basin, it may be rewarding to study in more detail the results obtained for Degerby, since this sea-level station, owing to its central position only slightly to the north of the middle part of the entire basin of the sea and its large gulfs, is a good representative of the average conditions in the whole area (Lisitzin, 1953). Concerning the day-to-day variations a certain inaccuracy may however be involved, if the results for Degerby are taken to be representative of the entire Baltic Sea. The changes in sea level at an individual station are without doubt greater than those for the whole sea basin. A comparative study indicated that a reduction amounting to 20% of the values obtained for Degerby is reasonable in order to provide acceptable values for the entire Baltic. According to Table LXVII the sum of the increase in sea level at Degerby amounts to 600 cm per year. Starting from this value the average increase for the whole sea basin may therefore be estimated at 480 cm per year. Taking into account that the average depth of the Baltic is approximately 60 m, this value implies that as an initial estimate somewhat less than 13 years should be required for a more-or-less 'complete' renewal of the water masses in the Baltic basin. In order to compare this period with the renewal times for other sea basins, according to Libby (1952) a similar period may be estimated for the Mediterranean Sea at 80 years and for the Arctic Ocean at 165 years. These periods are considerably more prolonged than that computed for the Baltic, but it must be borne in mind that the volume of the Mediterranean is 184 times, and that of the Arctic Ocean 938 times larger than that of the Baltic Sea. The renewal period of approximately 13 years seems therefore to be reasonable, although the considerable differences in the structure of the transition areas of the particular basins do not really allow general conclusions to be drawn.

Since the total area of the Baltic, including the extensive gulfs, is 365,000 km^2, the average amount of water participating every year in the renewal may be determined at 1,754 km^3. This water volume is almost twice as large as the water outflow determined for the Baltic basin by Brogmus (1952) according to the following equation which represents the annual water budget:

Run-off + precipitation $-$ evaporation + inflow   = outflow
472 km^3 + 172 km^3 $-$ 172 km^3 + 472 km^3 = 944 km^3

It may be of interest to mention in this connection that the first three terms on the left were determined by Brogmus with great care using all available observations, while the amounts of in- and outflow were computed on the basis of data for the mean salinity of the surface and bottom currents in the transition area. The starting point for these computations was the general assumption that the quantities of water and salt transported by the two currents represented a state of equilibrium with no changes in the total amount of water and salt. This procedure signifies that the inaccuracy inherent in the original data will be reflected in the values representing the in- and outflow.

Soskin and Rosova (1957) chose a different way of proceeding. They started with current observations in the transition area around Denmark and obtained for the years 1898–1944 the following results for water interchange between the North Sea and the Baltic:

Fresh water supply + inflow   = outflow
473 km^3 + 1187 km^3 = 1660 km^3

The results for water interchange determined on the basis of day-to-day variations in sea-level and completed for fresh water supply by the data computed by Witting (1918) may be reproduced in the following way:

Run-off   + precipitation − evaporation + inflow   = outflow
467 km^3 + 206 km^3 − 182 km^3 + 1263 km^3 = 1754 km^3

The total fresh water supply amounts in the last-mentioned case to 491 km^3. It may thus be established that the relevant data computed by three different authors using very different methods and periods deviate by less than 4%. These differences are therefore of slight significance for the result as a whole.

Recently Svansson (1972) determined the volume of the out- and ingoing water transport in the Baltic using 30-year means of the salinity determined for the light vessel 'Gedser Rev.' Svansson's results are:

Outgoing water − incoming water = difference
1718 km^3 − 1246 km^3 = 472 km^3

The results achieved on the basis of day-to-day variations of the sea level at Degerby are fairly close to the data reported by Soskin and Rosova and still closer to those determined by Svansson. Since in this case the determinations involved in computing the final result are based on completely different data the outcome is encouraging. The examination of water and salt interchange supports quite evidently the validity of the results based on sea-level data. The ratio between outflow and inflow is 1.39. This ratio is of the same magnitude as the ratios between the salinities of the bottom and surface currents in the transition area. Thomsen (1963), for instance, gave the average salinity values for the bottom and surface currents at the light vessel 'Gedser Rev,' the values used being more recent than those used by Svansson, and at 'Halskov Rev.' For the former light vessel these salinities were 14.6‰ and 9.7‰ respectively, and for the latter 22.2‰ and 15.2‰. The ratios concerned are thus 1.50 and 1.46. The correspondence between the three ratios is fairly pronounced. In the Öresund area the ratio between the salinity at the bottom and the surface may reach the value of two, or even higher values. Nevertheless, this fact cannot be considered as evidence against the results described above, since in this narrow transition region the ingoing current which causes a marked sea-level increase in the Baltic frequently reaches the surface.

Turning to the investigation of the total increase in sea level in the Gulfs of Bothnia and Finland, it may be noted that according to Table LXVII there is a total average yearly increase in sea level which at Kemi amounts to 21 m, at Raahe to more than 17.5 m and at Pietarsaari to 14 m. For the three stations situated in the Bothnian Bay, the average increase in sea level may thus be estimated at 17.5 m per annum. If we assume that this increase in sea level is also representative of the western parts of the Bothnian Bay, there is hardly any doubt that it must exercise a marked effect upon the exchange and renewal of water in this bay which has an average depth of approximately 42 m. The theoretical result is thus rather surprising: the water masses in the Bothnian Bay would be more or less renewed in a time-span which is less than 2.5 years.

In this connection there arises the interesting question of whether and to what extent the renewed water in the Bothnian Bay is water originating in the North Sea or whether it consists mainly of Baltic or possibly Bothnian Sea water. (The Bothnian Sea is the larger southern part of the Gulf of Bothnia, the Bothnian Bay its smaller northern part, see Fig. 46.) The answer can, of course, only be approximate. The time of the water trans-port from the Danish Straits to Utö, an island situated in the northern part of the Baltic proper, was determined by Lisitzin (1948), by means of salinity and sea-level data, at roughly 10 months. Ahlnäs (1962), basing her results on the effect of 5 large inflows of North Sea water in the Baltic and the salinity values at Utö, estimated that this time corresponds to slightly less than 8.5 months. Voipio (1964) computed from salinity data the mean velocity of the water transport in the Bothnian Sea and obtained a result of 4.2 cm per sec. If the velocity in the Bothnian Bay is assumed to be of the same magnitude, the time needed for the water masses to proceed from Utö to Kemi is 6.5 months. The time required to cover the whole distance from the transition area to the inner parts of the Bothnian Bay would, according to the above estimates, amount to 15–17 months. This estimate gives an indication that during the 2.5 years determined as a possible time for the renewal of the water in the Bothnian Bay, a certain amount of North Sea, or, to be more correct, Danish Strait water may, at least theoretically, reach the innermost parts of the Gulf of Bothnia.

In the Gulf of Bothnia as a whole the average increase in sea level may, according to Table LXVII, be estimated at 12.0 m per year. The average depth of the gulf being 60 m, a rough estimate is that a period of 5 years should be necessary for a theoretical renewal of the water in this gulf.

For the Gulf of Finland the renewal time is according to the estimates still shorter. Table LXVII shows that the mean sea-level increase per year in this gulf is 11.5 m. The average depth of the gulf is 38 m. The renewal of the water in the Gulf of Finland may therefore be expected to occur in a time covering slightly more than 3 years. The water renewal in the two extensive gulfs embracing Finland seems thus to take place during a comparatively short time. In this connection it must, however, be pointed out once more that it is inadequate to assume a total renewal of water. As a consequence of rapidly changing meteorological conditions, distinctly reflected in rapid and practically continu-ous sea-level fluctuations, the original water of the gulfs may return, at least partly, within a few hours or days. On the contrary, in specific regions where pollution of the water is highly pronounced — for instance as the result of an industrial plant situated close to the coast — rapid mixing with uncontaminated or only weakly contaminated water of the surrounding regions may be fairly efficacious.

The above results illustrating conditions in the Baltic basin show very distinctly the significance of sea-level data for the determination of the renewal time of the water in a basin. Since the determination of the renewal forms the basis for all investigations of water pollution in the coastal regions of all oceans and seas, more attention should be paid to sea-level variations in connection with environmental investigations in marine areas. It must particularly be borne in mind that sea-level data are as a rule highly reliable, and that they are available on a large scale.

Some additional problems closely connected with water pollution and marine environment studies, for the solution of which sea-level data may be of great use, should be mentioned. The use of the ocean floor as a receptacle for atomic wastes necessitates an exact knowledge of the tidal regime — of the tidal currents probably more than of the elevations in the deep oceanic regions. Unfortunately, there are gaps in this knowledge which must be eliminated in order to give a satisfactory picture of the tidal currents in the deep, open parts of the oceans. Although no definite results have so far been achieved, there are signs that the use of electronic computers will be fruitful for the solution of the problem of oceanic tides, at least if this problem is considered from a practical point of view. Moreover, it may be appropriate to mention the most exciting development in the field of tidal research in the open sea, the Vibroton. This instrument, originally developed for recording wave motion in shallow water, had had considerable success and is now being adopted both in Great Britain and in the United States to record tides and long waves in deep water. By supplementing the data provided by other deep-sea tide gauges it could possibly be the significant break-through in the investigations of off-shore tidal problems.

Corresponding to the importance of knowledge of extreme high sea levels and their occurrence for coastal protection, the abnormally low sea-level heights may be of considerable interest when studying various technical problems which have a close bearing on pollution of the water in coastal regions. It may be sufficient to mention in this respect the designs of water-cooling outfall channels for power stations, and other outlets. In these cases the frequency of the occurrence of low sea-level heights and their limits must be known.

APPENDIX

A FEW WORDS ABOUT PHENOMENA CONNECTED WITH SEA-LEVEL CHANGES
DURING THE PRE-CHRISTIAN ERA AND THEIR MODERN EXPLANATION

This survey is by no means intended to be complete. It is a shortened version of a lecture given before the Finnish Society of Sciences (Societas Scientiarum Fennica). The principal purpose is to give an idea of the significant role sea-level variations played in the life of ancient peoples (Lisitzin, 1961c).

Only a few natural phenomena have aroused since the dim and distant past such interest and attention as disastrous floods. This fact is hardly surprising when taking into account the fact that coastal regions have always been affected by devastating catastrophes of this kind. The causes of these catastrophes must have appeared to primitive man as mysterious as their consequences were disastrous. A considerable number of accounts concerning devastating floods which occurred many thousands of years ago have been preserved up to the present. Sumerians and Babylonians, Greeks and Jews, Indians and Arabs have left their testimony on this subject. These accounts should not be considered only as a people's endeavour to give their imagination a free rein. Without doubt, they must have been based on actual events, since they describe at least some of the phases of the phenomena in very natural manner. In addition, the consequences of the floods can in some cases be confirmed by archaeological excavations. On the other hand, it can hardly be denied that some of the accounts depicting floods are loans from one people to another neighbouring people. Let us mention a few examples. In an old Sumerian description of the flood, dating probably from the third millenium B.C., the following lines may be read: 'The strong wind arrived with an enormous power and the devastating cyclone roared ahead. When the cyclone had raced over the country for seven days and seven nights and the big ship had rocked around on the huge water, the Sun God Shamash appeared and his light began to shine over sky and land'. The Babylonian account of the deluge, which is a part of the extensive and famous heroic Epic of Gilgamesh, shows several features which could be interpreted as a loan, but it is considerably richer in detail. The style is extremely vivid. One almost hears the roaring of the storm, when one reads the following account. 'The hurricane raved uninterruptedly during the whole day, it advanced like a hurrying storm and forced the water waves up to the tops of the mountains, the storming flood rushed as on a battlefield over the people'. In comparison with this lifelike description the account of the deluge given in the Old Testament appears almost dull. It states quite briefly, that the flood reached the land and that the water rose more and more in such a way that all high mountains everywhere under the sky were covered.

The Austrian geologist Eduard Suess in 1883 made an attempt to reconstruct the main features of the deluge. He stated: 'The natural catastrophe known as the deluge occurred in the lower course of the Eufrates and was combined with an extensive, devastating flood in the Mesopotamian lowlands. The principal cause of the event was a strong submarine earthquake in the Persian Gulf, or in the area situated to the south of it. It had been preceded by a series of minor earthquakes. It is possible that a cyclone broke over Mesopotamia from the Persian Gulf during the occurrence of the strongest quake'.

In connection with the excavations on the site of the old biblical town of Ur in Mesopotamia, the English archaeologist Leonard Woolley found traces of an exceptionally strong flood, and it is therefore quite reasonable to assume that this flood was identical with the deluge recounted in the Old Testament, and possibly also with its Sumerian and Babylonian forerunners. Woolley found in Ur a clay layer which was 600 km long and 150 km broad and which was an obvious proof that this area at some times in the very distant past was reached by a flood with an estimated height of about 7 m. An increase in sea-level of such dimensions is, without doubt, sufficient to cause a tremendous disaster. It may, however, be pointed out, that only in a completely flat country lying slightly above the average level of the sea could the flood have had such devastating consequences as those mentioned in the Old Testament.

The occurrence of another pre-Christian flood can be proved by excavations on the volcanic island of Thera in the Aegean Sea. These excavations have shown very distinctly that the volcano on Thera, approximately in the year 1500 B.C., must have been the cause of a veritable catastrophe. It is by no means impossible that the Greek myth about the rescue of the son of Prometheus, Deukalion, and his wife in a tiny boat originated as a consequence of the enormous flood waves which reached the mainland of Greece from Thera.

Disastrous floods are doubtless the most dramatic chapter in the investigations of sea-level changes, but fortunately they do not belong to everyday phenomena. On the contrary, the tide is one of the most regular phenomena in nature. In spite of the pronounced difference in their character floods and tides have one feature in common; even tides have an interesting historical or, let us say, literary background.

About 800 B.C. the great Greek poet Homer referred in the Odyssey to a phenomenon which must be connected with the tide. He vividly described two terrible sea monsters, Scylla and Charybdis, watching a narrow sound and obstructing the navigation. The myth of these sea monsters survived during many centuries and Virgil also mentioned the danger they brought about to navigators.

Modern oceanography considers it highly probable that the sound described by Homer is identical with the Strait of Messina, situated between the mainland of Italy and Sicily. It is a commonly known fact that the tides have different phases in the Tyrrhenian Sea to the north of the strait and the Ionian Sea to the south of it. The consequence of this fact is that the sea level may be high at the northern entrance to the strait and low at the southern approaches, and vice versa. These height differences in sea level are the cause of the extremely strong currents which may occur in the strait. These currents change their direction approximately every six hours. Winds blowing from certain directions may contribute to the increase in velocity of these currents. In addition, the water in the Ionian Sea is not only colder but also salt-richer than the water in the Tyrrhenian Sea and consequently heavier. When penetrating underneath the lighter Tyrrhenian water, it may cause fairly powerful water eddies which in some special situations favourable for their occurrence may reach considerable dimensions. These eddies are generally located in the narrowest parts of the Strait of Messina. A weaker eddy may also occur at the northern entrance to the sound, at a place which still today bears the name 'Scilla'. Although this eddy is at the present time of relatively slight significance, this was by no means the case during ancient times. A powerful earthquake which occurred in 1783, according to the Italian scientist Mazzarelli, dislocated enormous rocks outside Scilla with the consequence that the large caverns which before the earthquake opened at the level of the sea surface have disappeared. It is highly probable that these caverns, when filled by racing water masses, produced the different sounds so vividly described by Homer. Some oceanographers are, moreover, of the opinion that the Strait of Messina itself was in former days not only narrower, but also shallower than it is now. The currents and eddies were therefore more strongly developed in these long-bygone days and so corresponded better to the fantastic descriptions given by Homer and later by Virgil.

It may appear rather surprising to us that the ancient Greeks, who, as a rule, had a considerable capacity to observe natural phenomena and generally tried to explain their origin also, failed to obtain knowledge concerning the tides. This deficiency was probably related to the fact that the tides are weak within practically the whole Mediterranean area, the differences in height between high water and low water amounting, as a rule, only to a few centimetres. However, it may be mentioned that the first-known Greek historian, Herodotus, who lived from approximately 484–424 B.C., knew of the tide which occurs in the Red Sea.

About 330 years before the beginning of the Christian era, an erudite man called Pytheas from the Greek colony Massilia, a forerunner of modern Marseilles, started on a long maritime journey which covered the western parts of the Mediterranean and the Atlantic Ocean and reached the distant British Isles. On arrival at the last-named, Pytheas was highly astonished to observe the impressive sea-level variations caused by the tide. Today it is a well-known fact that in some parts of the English Channel the tides are among the most pronounced in Europe. The regular fluctuations in sea level and their connection with the changing phases of the Moon could hardly escape the attention of Pytheas, particularly since, as an astronomer, he was interested in celestial bodies. Unfortunately Pytheas' manuscripts have long ago disappeared, but on the basis of quotations made by Strabo and Pliny it seems fair to consider Pytheas as the first person, or, let us say, the first scientist who observed the close connection between the varying phases of the Moon and the regular periodic changes in the rhythm of the tide.

Not only itinerant navigators such as Odysseus and navigating astronomers such as Pytheas encountered the phenomenon of the tide in connection with their travels during ancient times. World-famous strategists were also unable to escape it. Various historical authors have mentioned the amazement of Alexander the Great and his army, when during the illustrious campaign to India in 325 B.C. they experienced for the first time in their lives sea-level variations caused by tide and noted the disastrous consequences of this phenomenon on the Macedonian fleet, anchored in the estuary of the River Indus.

Julius Caesar also met with defeat caused by tidal fluctuations between extremely high and low water. In 55 B.C., some 300 years after Alexander the Great's campaign in India, Caesar tried to cross the English Channel with his fleet. In *De Bello Gallico* Julius Caesar mentioned that it was full moon during the night of invasion, an epoch when the tidal variations are at their largest, although this fact was not known to him and his men. Caesar also described the considerable damage he suffered during the expedition. The military result of Caesar's first attack on the British Isles was therefore rather slight, principally because he was not able to ship over his cavalry.

In this connection it may be of interest to draw a parallel between the difficulties which Julius Caesar encountered during his first endeavour to subdue the British Isles and the invasion of the Allied forces in Normandy during the Second World War. The most significant difference from the standpoint of sea-level scientists is self-evidently not the more modern and better equipment which was at the disposal of the Allied forces, but the fact that an accurate forecast of the tidal sea-level variations could be made for every locality along the French coast of the English Channel and for every period in time. The headquarters of the Allied forces had therefore no difficulty in deciding beforehand which day and hour were the most favourable for the operation.

Time changes and the methods for sea-level research develop rapidly, but the phenomena themselves always remain unchanged.

REFERENCES

Ahlnäs, K., 1962. Variations in salinity at Utö 1911–1961. *Geophysica*, 8: 1–16.

Aliverti, G., Picotti, M., Trotti, L., De Maio, A., Lauretta, O. and Moretti, M., 1968. *Atlante del Mar Tirreno, Isoterme ed Isohaline.* Istituto Universitario Navale, Napoli, 127 pp.

Angeby, O., 1953. Stormfloden i Holland den 1 Febr. 1953. *Svensk Geogr. Årsbok*, 29: 141–152.

Anonymous, 1963. *Handbuch für das Rote Meer und den Golf von Aden.* Deutsches Hydrographisches Institut, Hamburg, 569 pp.

Association d'Océanographie Physique, 1940. Monthly and annual heights of sea level up to and including the year 1936. *Publ. Sci.*, 5: 255 pp.

Association d'Océanographie Physique, 1950. Monthly and annual mean heights of sea level 1937–1946 and unpublished data for earlier years. *Publ. Sci.*, 10: 82 pp.

Association d'Océanographie Physique, 1953. Monthly and annual mean heights of sea level 1947 to 1951 and unpublished data for earlier years. *Publ. Sci.*, 12: 61 pp.

Association d'Océanographie Physique, 1955. Bibliography on tides 1665–1939. *Publ. Sci.*, 15: 220 pp.

Association d'Océanographie Physique, 1957a. Bibliography on tides 1940–1954. *Publ. Sci.*, 17: 63 pp.

Association d'Océanographie Physique, 1957b. Bibliography on generation of currents and changes of surface level in oceans, seas and lakes by wind and atmospheric pressure 1726–1955. *Publ. Sci.*, 18: 83 pp.

Association d'Océanographie Physique, 1958. Monthly and annual mean heights of sea level 1952–1956 and unpublished data for earlier years. *Publ. Sci.*, 19: 78 pp.

Association d'Océanographie Physique, 1959. Monthly and annual mean heights of sea level for the period of the International Geophysical Year (1957–1958) and unpublished data for earlier years. *Publ. Sci.*, 20: 65 pp.

Association d'Océanographie Physique, 1963. Monthly and annual mean heights of sea level 1959 to 1961 and unpublished data for earlier years. *Publ. Sci.*, 24: 59 pp.

Association d'Océanographie Physique, 1964. Bibliography on mean sea level 1719–1958. *Publ. Sci.*, 25: 25 pp.

Association Internationale des Sciences Physiques des Océans, 1968. Monthly and annual mean heights of sea level 1962 to 1964. *Publ. Sci.*, 26: 109 pp.

Association Internationale d'Océanographie Physique, 1971a. Bibliography on mean sea level 1959–1969. *Publ. Sci.*, 29: 4–18.

Association Internationale d'Océanographie Physique, 1971b. Bibliography on tides 1955–1969. *Publ. Sci.*, 29: 19–40.

Avers, H.G., 1927. A study of the variation of mean sea level from a level surface. *U.S. Coast Geodetic Surv., Spec. Publ.*, 134: 28–30.

Baussan, J., 1951. La composante de Chandler dans la variation des niveaux marins. *Ann. Géophys.*, 7: 59–62.

Benton, G.S. and Estoque, M.A., 1954. Water vapour transfer over the North American continent. *J. Meteorol.*, 11: 462–477.

Bergsten, F., 1930. The changes of land-level at the Swedish coasts computed with regard to periodic fluctuations of sea level. *3e Conf. Hydrol. Etats Baltique*, 27: 1–6.

Blomqvist, E. and Renqvist, H., 1914. Wasserstandsbeobachtungen an den Küsten Finnlands. *Fennia*, 37: 1–433.

Borre, K., 1970. The influence of current and meteorological forces on the mean sea level in the Danish Straits. *Geod. Inst. Medd. Kbh.*, 47: 63 pp.

Bowden, K.F., 1956. The flow of water through the Straits of Dover related to wind and differences in sea level. *Philos. Trans. R. Soc., A*, 248: 517–551.

Bowden, K.F., 1960. The effect of water density on the mean slope of the sea surface. *Bull. Géod.*, N.S., 55: 93–96.

Braaten, N.F. and McCombs, C.E., 1963. *Mean Sea Level Variations as Indicated by a 1963 Adjustment of First-Order Leveling in the United States.* Coast and Geodetic Survey, Washington, D.C., 22 pp.

Brogmus, W., 1952. Eine Revision des Wasserhaushaltes der Ostsee. *Kieler Meeresforsch.*, 9: 15–42.

Budyko, M.I., 1956, *The Heat Balance of the Earth's Surface.* (Translated by N.A. Stepanova, 1958). U.S. Weather Bureau, Washington D.C., 257 pp.

Caloi, P., 1938. Sesse dell'Alto Adriatico con parto, riguardo la Golfo di Trieste. *Mem. R. Com. talassogr. Ital.*, 247, 39 pp.

Carruthers, J.N. and Lawford, A.L., 1954. The progress of the storm surge of January 31st, February 1st, 1953, as indicated by current meter observations in the southern North Sea. *Assoc. Int. Océanogr. Phys., Proc. Verb.*, 6: 197–198.

Cartwright, D.E., 1968. A unified analysis of tides and surges round north and east Britain. *Philos. Trans. R. Soc., A.*, 263: 1–55.

Cartwright, D.E., 1969. Deep-sea tides. *Sci. J.*, 5: 60–67.

Cartwright, D.E., Munk, W. and Zetler, B., 1969. Pelagic tidal measurements *Trans. Am. Geophys. Union*, 50: 472–477.

Celsius, A., 1743. Anmärkning om vatnets förminskande, så i Östersjön som Vesterhafvet. *K. Svenska Vetensk. Akad. Handl.*, 4: 33–50.

Chrystal, G., 1905. On the hydrodynamic theory of seiches. *Trans. R. Soc., Edinb.*, 41: 599–649.

Cochrane, J.D., 1958. The frequency distribution of water characteristics in the Pacific Ocean. *Deep-Sea Res.*, 5: 111–127.

Colding, A., 1880. Resultaterne af nogle Undersøgelser over de ved Vindens Kraft framkaldte Strømninger i Havet. *Danske Vid. Sels. Skrift. Natur-Math.*, 11: 247–274.

Corkan, R.H., 1948. *Storm Surges in the North Sea.* U.S. Hydrogr. Off., Misc. 15072, Vol. 1, 174 pp. and Vol. 2, 166 pp.

Crease, J., 1956. Propagation of long waves due to atmospheric disturbances on a rotating sea. *Proc. R. Soc., A*, 233: 556–569.

Darwin, G.H., 1898. *The Tides and Kindred Phenomena in the Solar System.* Boston, Mass., and London, 342 pp.

Deacon, G.E.R., 1937. Note on the dynamics of the Southern Oceans. *"Discovery" Rep.*, 15: 125–152.

Defant. A., 1918. Neue Methode zur Ermittlung der Eigenschwingungen (Seiches) von abgeschlossenen Wassermassen. *Ann. Hydrog. Marit. Meteorol.*, 46: 78–85.

Defant, A., 1919. Untersuchungen über die Gezeitenerscheinungen in Mittel- und Randmeeren, in Buchten and Kanälen. *Denkschr. Akad. Wiss., Wien*, 96: 110–137.

Defant, A., 1926. Gezeiten und Gezeitenströmungen im Roten Meer. *Ann. Hydrogr. Marit. Meteorol.*, 54: 185–194.

Defant, A., 1940. Scylla und Charybdis und die Gezeitenströmungen in die Strasse von Messina. *Ann. Hydrogr. Marit Meteorol.*, 68: 145–157.

Defant, A., 1941. Quantitative Untersuchungen zur Statik und Dynamik des Atlantischen Ozeans. 5. Lief. Die absolute Topographie des physikalischen Meeresniveaus und der Druckflächen, sowie die Wasserbewegung in Atlantischen Ozean. *Wissensch. Ergebn. Dtsch. Atlant. Exped. "Meteor"*, 6: 191–260.

Defant, A., 1961. *Physical Oceanography.* Pergamon, Oxford and New York, N.Y., Vol. 1, 729 pp.; Vol. 2, 598 pp.

Dietrich, G., 1935. Aufbau und Dynamik des Südlichen Agulhasstromgebietes. *Veroff. Inst. Meeresk. Berlin, N.F. Geogr-Naturw. Reihe*, 27: 3–79.

Dietrich, G., 1937. Die Lage der Meeresoberfläche im Druckfeld von Ozean und Atmosphere, mit besonderer Berücksichtigung des westlichen Nordatlantischen Ozeans und des Golfs von Mexico. II. Über Bewegung und Herkunft des Golfstromwassers. *Veröff. Inst. Meeresk. Berlin, N.F. Geogr.-Naturw. Reihe*, 33: 1–52.

Dietrich, G., 1944. Die Gezeiten des Weltmeeres als geographische Erscheinung. *Z. Ges. Erdkd. Berlin*, 3/4: 69–85.

Dietrich, G., 1952. Physikalische Eigenschaften des Meerwassers. In: Landolt-Börnstein: *Zahlenwerte und Funktionen*, 3. Springer, Heidelberg, pp. 426–441.

Dietrich, G., 1954. Ozeanographisch-meteorologische Einflüsse auf Wasserstandsänderungen des Meeres am Beispiel der Pegelbeobachtungen von Esbjerg. *Die Küste*, 2: 130–156.

Dietrich, G., 1963. *General Oceanography*. Wiley, New York, N.Y., 588 pp.

Dines, J.S., 1929. Meteorological conditions associated with high tides in the Thames. *Geophys. Mem. London*, 47: 27–39.

Disney, L.P., 1954. Report on the investigation of the secular variation of sea level along the coast of America, the Hawaiian and Philippine Islands, and Japan. *Assoc. Océanogr. Phys., Publ. Sci.*, 13 *(Secular Variation of Sea-Level)*: 11–15.

Disney, L.P., 1955. Tide heights along the coasts of the United States. *Proc. Amer. Soc. Civ. Engineers*, 81 (666): 1–9.

Doodson, A.T., 1921. The harmonic development of the tide-generating potential. *Proc. R. Soc., A*, 100: 305–329.

Doodson, A.T., 1956. Tides and storm surges in a long uniform gulf. *Proc. R. Soc., A.*, 237: 325–345.

Doodson, A.T., 1960. Mean sea level and geodesy. *Bull. Géod.*, 55: 69–88.

Duvanin, A.I., 1956. *The Sea Level*. Gidrometeorologicheskoie Izdatelstwo, Leningrad, 60 pp. (in Russian).

Egedal, J., 1954. Report on the investigation of the secular variation of sea level on the coasts of Europe (except the British Isles) and of North Africa. *Assoc. Océanogr. Phys., Publ. Sci.*, 13 *(Secular Variation of Sea-Level)*: 4–10.

Ekman, V.W., 1906. Beiträge zur Theorie des Wasserströmungen. *Ann. Hyrogr. Marit. Meteorol.*, 34: 423–430, 472–484, 527–540, 566–583.

Ertel, H., 1933. Eine neue Methode zur Berechnung der Eigenschwingungen von Wassermassen in Seen unregelmässiger Gestalt. *Sitz. Ber. Preuss. Akad. Wiss.*, 24: 746–753.

Exner, F.M., 1925. *Dynamische Meteorologie*. Springer, Wien, 421 pp.

Eyriès, M., 1968. Marégraphes de grandes profondeurs. *Cah. Océanogr. C.O.E.C.*, 20: 355–368.

Fairbridge, R.W., 1961. Eustatic changes in sea-level. In: *Physics and Chemistry of the Earth*. Pergamon, Oxford and New York, N.Y., 4: 99–185.

Fedorov, K.N., 1959. The causes of the semi-annual periodicity in atmospheric and oceanic processes. *Izv. Akad. Nauk SSSR, Ser. Geograf.*, 4: 17–25 (in Russian).

Filloux, J.H., 1968. Deep-sea tide record from northeastern Pacific (Abstract). *Trans. Am. Geophys. Union*, 49: 211.

Fischer, G., 1959. Ein numerisches Verfahren zur Errechnung von Windstau und Gezeiten in Randmeeren. *Tellus*, 11: 60–76.

Forel, F.A., 1895. *Le Léman*, 2. Libraire de l'Université, Lausanne, 651 pp.

Frasetto, R., 1970. The subsidence and storm surge effects in Venice – Italy. *Rep. Symp. Coastal Geodesy, Munich*, pp. 527–535.

Fuglister, F.C., 1951. Annual variations in current speed in the Gulf Stream system. *J. Mar. Res.*, 10: 119–127.

Galerkin, L.I., 1960. On the physical basis of the forecast of the seasonal variations of sea level in the Sea of Japan. *Tr. Inst. Okeanol. SSSR*, 37: 73–91 (in Russian).

Galerkin, L.I., 1963. The static effect of atmospheric pressure upon the seasonal sea level variations in the Pacific Ocean. *Okeanologia*, 3: 384–394 (in Russian).

Galerkin, L.I., Shagin, V.A. and Nefediev, V.P., 1962. The seasonal variations in sea level in the Austro-Asiatic Seas. *Tr. Inst. Okeanol. SSSR*, 60: 161–177 (in Russian).

Gill, A.E. and Niiler, P.P., 1973. The theory of the seasonal variability in the ocean. *Deep-Sea Res.*, 20: 141–177.

Godin, G., 1972. *The Analysis of Tides*. Liverpool University Press, Liverpool, 272 pp.

Grace, S.F., 1930. The semi-diurnal lunar tidal motion of the Red Sea. *Mon. Not. R. Astr. Soc., Geophys. Suppl.*, 2: 273–296.

Granqvist, G., 1938. Zur Kenntnis der Temperatur und des Salzgehaltes des Baltischen Meeres an den Küsten Finnlands. *Merentutkimuslaitoksen Julkaisu-Havsforskningsinstitutets Skrift*, 122: 166 pp.

Groen, P., 1953. Voorlopig onderzoek van de waterstanden opgetreden langs de kusten der Noordzee op 31 Jan. en 1 Febr. 1953. *K. Ned. Meteorol. Inst. Rapp.*, 6: 5 pp.

Groen, P., 1954. Analyse van het verloop der waterstanden langs de Kusten der Noordzee tijdens de stormvloed van 1953. *K. Ned. Meteorol. Inst. Rapp.*, 1 (vervolg): 7–11.

Groen, P. and Groves, G.W., 1962. Surges. In: M.N. Hill (Editor), *The Sea*. Interscience, New York, N.Y. and London, 1: 611–646.

Guilcher, A., 1953. Les inondations marines du 31 janvier et du 1er février 1953 sur les bords de la mer du Nord. *Bull. Union Int. Secours*, 12: 3–19.

Gutenberg, B., 1941. Changes in sea level, postglacial uplift and mobility of the earth's interior. *Geol. Soc. Am. Bull.*, 52: 721–772.

Hankimo, J., 1964. Some computations of the energy exchange between the sea and the atmosphere in the Baltic area. *Finn. Meteorol. Off. Contrib.*, 57: 3–26.

Hansen, W., 1948. Neuere Ergebnisse der Gezeitenforschung. *Naturwissenschaften*, 35: 265–269.

Hansen, W., 1949. Die halbtägigen Gezeiten im Nordatlantischen Ozean. *Dtsch. Hydrogr. Z.*, 2: 44–51.

Hansen, W., 1952a. Gezeiten in beliebig gestalteten Meeresgebieten. In: Landolt-Börnstein: *Zahlenwerte und Funktionen*, 3. Springer, Heidelberg, pp. 521–526.

Hansen, W., 1952b. Gezeiten und Gezeitenströme der halbtägigen Hauptmondtide M_2 in der Nordsee. *Dtsch. Hydrogr. Z., Erg. Heft*, 1: 46 pp.

Hansen, W., 1956. Theorie zur Errechnung des Wasserstandes und der Strömungen in Randmeeren nebst Anwendungen. *Tellus*, 8: 287–300.

Hansen, W., 1962. Tides. In: M.N. Hill (Editor), *The Sea*. Interscience, New York, N.Y. and London, pp. 764–801.

Harris, R.A., 1904. *Manual of Tides, Part IVb*. Coast and Geodetic Survey Report, Washington, D.C., pp. 313–400.

Hatori, T., 1963. Directivity of tsunamis. *Bull. Earthq. Res. Inst.*, 41: 61–81.

Haubrich, R. and Munk, W.H., 1959. The pole tide. *J. Geophys. Res.*, 64: 2373–2388.

Heaps, N.S., 1967. Storm surges. *Oceanogr. Mar. Biol., Ann. Rev.*, 5: 11–47.

Heck, N.H., 1947. List of seismic sea waves. *Bull. Seism. Soc. Am.*, 37: 269–286.

Hela, I., 1944. Über die Schwankungen des Wasserstandes in der Ostsee mit besonderer Berücksichtigung des Wasseraustausches durch die dänischen Gewässer. *Merentutkimuslaitoksen Julkaisu-Havsforskningsinstitutets Skrift*, 134: 108 pp.

Hela, I., 1952. The fluctuations of the Florida current. *Bull. Mar. Sci.*, 1: 241–248.

Hela, I., 1953. A study of land upheaval at the Finnish coast. *Fennia*, 76: 38 pp.

Hela, I., 1957. Longitudinal and transversal slope of the Florida current. *Geophysica*, 5: 1–8.

Hela, I., 1960. The hydrographical features of the Baltic Sea and the disposal of radioactive wastes. In: *Disposal of radioactive wastes*. IAEA, Vienna, pp. 573–587.

Hellström, B., 1941. Wind effect in lakes and rivers. *Handl. Ing. Vetensk. Akad.*, 158: 191 pp.

Heyer, E. and Grünewald, G., 1953. Der Nordseeorkan vom 31. Januar bis 1. Februar 1953 und seine Ursachen. *Z. Meteorol.*, 7: 176–183.

Hicks, S.D., 1967. The tide prediction centenary of the United States Coast and Geodetic Survey. *Int. Hydrogr. Rev.*, 44: 121–131.

Hicks, S.D., 1968. Long period variations in secular sea level trends. *Shore Beach.*, 36: 32–36.

Hicks, S.D., 1972a. Changes in tidal characteristics and tidal datum planes. In: *The Great Alaska Earthquake of 1964. Oceanography and Coastal Engineering*. National Academy of Sciences, Washington, D.C., pp. 310–314.

Hicks, S.D., 1972b. On the classification and trends of long period sea level series. *Shore Beach*, 40: 20–23.

Hicks, S.D., 1973. Trends and variability of the yearly mean sea level 1893–1971. *NOAA Tech. Memo. NOS 12*, U.S. Department of Commerce, Rockville, Md., 13 pp.

Hicks, S.D. and Shofnos, W., 1965. Yearly sea level variations for the United States. *J. Hydraul. Div. Amer. Soc. Civ. Engineers*, 91: 23–32.

Hidaka, K., 1936. Application of Ritz's variation method to the determination of seiches in a lake. *Mem. Kobe Marine Obs.*, 6: 159–174.

Iida, K., 1963. Magnitude, energy and generation mechanisms of tsunamis and catalogue of earth-quakes associated with tsunamis. In: *Proc. Tsunami Meet. associated with the Tenth Pac. Science Congress, IUGG Monographe*, 24: 7–18.

Irish, J., Munk, W. and Snodgrass, F. 1971. M_2 amphidrome in the Northeast Pacific. *Geophys. Fluid Dyn.*, 2: 355–360.

Iselin, C.O'D., 1940. Preliminary report on long-period variations in the transport of the Gulf Stream system. *Pap. Phys. Oceanogr.*, 8: 1–40.

Jacobs, W.C., 1951. The energy exchange between the sea and atmosphere and some of its conse-quences. *Bull. Scripps Inst. Oceanogr.*, 2: 27–122.

Jacobsen, J.P., 1943. The Atlantic current through the Faroe–Shetland Channel and its influence on the hydrographical conditions in the northern part of the North Sea, the Norwegian Sea and the Barents Sea. *Cons. Perm. Int. Explor. Mer., Rapp. Proc. Verb. Réun.*, 122: 5–47.

Jakubovsky, O., 1966. Vertical movements of the earth's crust on the coasts of the Baltic Sea. *Ann. Acad. Sci. Fennicae, A III*, 90: 479–488.

Jeffreys, H., 1952. Tidal friction. In: *The Earth – Its Origin, History and Physical Constitution* (3rd ed.). Cambridge Univ. Press, Cambridge, Mass., pp. 217–248.

Jessen, A., 1955. Nivellement hydrographique entre six stations danoises. *Tellus*, 7: 381–384.

Jessen, A., 1964. Chandler's period in the mean sea level. *Tellus*, 16: 513–516.

Kääriäinen, E., 1966. The second levelling of Finland in 1935–1955. *Veröff. Finn. Geod. Inst.*, 61: 313 pp.

Karklin, V.P., 1967. The semi-annual variations in sea level in the Atlantic Ocean. *Okeanologia*, 7: 987–996 (in Russian).

Keuning, H.J., 1953. Die Stormflut vom 1 Februar in Niederland und ihre wirtschaftlichen Auswir-kungen. *Die Erde*, 3/4: 208–223.

King, C.A.N., 1962. *Oceanography for Geographers*. Edward Arnold, London, 337 pp.

Krauss, W. and Magaard, L. 1961. Zum Spektrum der internen Wellen der Ostsee. *Kieler Meeresforsch.*, 17: 137–147.

Krauss, W. and Magaard, L., 1962. Zum System der Eigenschwingungen der Ostsee. *Kieler Meeresforsch.*, 18: 184–186.

Kuenen, Ph.H., 1950. *Marine Geology*. Wiley, New York, N.Y., 551 pp.

Lacombe, H., 1951. Application de la méthode dynamique à la circulation dans l'océan Indien, au printemps boréal, et dans l'océan Antarctique, pendant l'été austral. *Bull. Inf. C.O.E.C.*, 3: 459–468.

Lacombe, H., 1959. Quelques reflexions sur le niveau moyen des mers. *C. R.Comm. Natn. Géod. Géophys.*, 59: 29–41.

La Fond, E.C., 1939. Variations of sea level on the Pacific Coast of the United States. *J. Mar. Res.*, 2: 17–29.

Laska, M., 1968. Wahania poziomu wod Baltyku oras hydrodynamiczno-numeryczna metoda ich obliczeń. (Summary: Water level changes in the Baltic and their calculations by means of hydro-dynamical-numerical method). *Inst. Budownictwa Wodnego Polskiej Akad. Nauk, Rozpr. Hydrotech.*, 22: 171–198.

Laska, M., 1969. Spectral analysis of the periodic water level changes in the Baltic. *Int. Hydrogr. Rev.*, 46: 115–129.

Lauwerier, H.A., 1956a. The influence of the disturbance upon an infinitely large shallow sea of constant depth. *Rapp. Afd. Toegep. Wisk. Math. Cent.*, TW 35: 11 pp.

Lauwerier, H.A., 1956b. The wind effect in the southern part of the North Sea due to a single storm and the influences of the Channel. *Rapp. Afd. Toegep. Wisk. Math. Cent.*, TW 36: 16 pp.

Lauwerier, H.A. and Damsté, B.R., 1963. The North Sea Problem, VIII. A numerical treatment. *Proc. K. Ned. Akad. Wet., A*, 66: 167–184.

Lennon, G.W., 1963. A frequency investigation of abnormally high tidal levels at certain west coast ports. *Proc. Inst. Civ. Engineers*, 25: 451–484.

Levallois, J.J. and Maillard, J., 1970. The new French 1st order levelling net – practical and scientific consequences. *Rep. Symp. Coastal Geodesy, Munich*, pp. 300–330.

Libby, F.W., 1952. *Radiocarbon Dating.* University of Chicago Press, Chicago, III., p. 25.

Lisitzin, E., 1944. Die Gezeiten des Finnischen Meerbusens, *Fennia*, 68: 19 pp.

Lisitzin, E., 1948. On the salinity in the northern part of the Baltic. *Fennia*, 70: 24 pp.

Lisitzin, E., 1952. Contribution to the knowledge of the range of the sea level variation in the North Baltic. *Merentutkimuslaitoksen Julkaisu- Havsforskningsinstitutets Skrift*, 153: 20 pp.

Lisitzin, E., 1953. Les variations du niveau de la mer dans la Baltique. *Bull. Inf. C.O.E.C.,* 5: 25–29.

Lisitzin, E., 1954a. Contribution to the knowledge of the annual sea level variations in the Northern Baltic. *Merentutkimuslaitoksen Julkaisu- Havsforskningsinstitutets Skrift*, 164: 12 pp.

Lisitzin, E., 1954b. Les variations du niveau de la mer à Monaco; Comparaison avec quelques autres stations marégraphiques de le côte française et italienne. *Bull. Inst. Océanogr. Monaco*, 1040: 24 pp.

Lisitzin, E., 1955a. Contribution à la connaissance des courants dans la mer Ligure et la mer Tyrrhénienne. *Bull. Inst. Océanogr. Monaco*, 1060: 11 pp.

Lisitzin, E., 1955b. La fréquence des différents niveaux mensuels de la mer à Brest. *Bull. Inf. C.O.E.C.*, 7: 407–410.

Lisitzin, E., 1955c. Les variations annuelles du niveau des océans. *Bull. Inf. C.O.E.C.*, 7: 235–250.

Lisitzin, E., 1956. Les variations semi-annuelles du niveau de la mer dans les océans. *Bull. Inf. C.O.E.C.*, 8: 343–353.

Lisitzin, E., 1957a. On the reducing influence of the sea ice on the pilling-up of water due to wind stress. *Comment. Phys.–Math. Helsingf.*, 20: 12 pp.

Lisitzin, E., 1957b. The annual variation of the slope of the water surface in the Gulf of Bothnia. *Comment. Phys–Math. Helsingf.*, 20: 20 pp.

Lisitzin, E., 1957c. The frequency of extreme heights of sea level along the Finnish coast. *Merentutkimuslaitoksen Jukaisu-Havsforskningsinstitutets Skrift*, 175: 12 pp.

Lisitzin, E., 1957d. The tidal cycle of 18.6 years in the oceans. *J. Cons. Int. Explor. Mer.*, 22: 147–151.

Lisitzin, E., 1958a. Determination of the slope of the water surface in the Gulf of Finland. *Geophysica*, 5: 193–202.

Lisitzin, E., 1958b. Le niveau moyen de la mer. *Bull. Inf. C.O.E.C.*, 10: 254–262.

Lisitzin, E., 1959a. Les variations saisonnières du niveau de la mer et de la densité de l'eau en Méditerranée Occidentale. *Cah. Océanogr. C.O.E.C.*, 11: 7–12.

Lisitzin, E., 1959b. The frequency distribution of sea-level heights along the Finnish coast. *Merentutkimuslaitoksen Julkaisu-Havsforskningsinstitutets Skrift*, 190: 37 pp.

Lisitzin, E., 1959c. The influence of water density variations on sea level in the Northern Baltic. *Int. Hydrogr. Rev.*, 36: 154–159.

Lisitzin, E., 1959d. Uninodal seiches in the oscillation system Baltic proper – Gulf of Finland. *Tellus*, 4: 459–466.

Lisitzin, E., 1960. L'effet de la pression atmosphérique sur les variations du niveau des océans. *Cah. Océanogr. C.O.E.C.*, 12: 461–466.

Lisitzin, E., 1961a. Les variations saisonnières du niveau de l'Océan Glacial Arctique. *Cah. Océanogr. C.O.E.C.*, 13: 161–166.

Lisitzin, E., 1961b. The effect of air pressure upon the seasonal cycle of sea-level in the oceans. *Comment. Phys-Math. Helsingf.*, 26: 19 pp.

Lisitzin, E., 1961c. Vattenståndsforskning och dess historiska bakgrund. *Soc. Sci. Fennica, Årsbok*, 39 B: 16 pp.

Lisitzin, E., 1962. La déclivité de la surface de la mer dans la Baltique. *Cah. Océanogr. C.O.E.C.*, 14: 391–397.

Lisitzin, E., 1963. Mean sea level. *Océanogr. Mar. Biol., Ann. Rev.*, 1: 27–45.

Lisitzin, E., 1964a. Contribution to the knowledge of land uplift along the Finnish coast. *Fennia*, 89: 22 pp.

Lisitzin, E., 1964b. La pression atmosphérique comme cause primaire des processus dynamiques dans les océans. *Cah. Océanogr. C.O.E.C.*, 16: 1–6.

Lisitzin, E., 1964c. Les causes des variations saisonnières du niveau de l'Océan Arctique. *Cah. Océanogr. C.O.E.C.*, 16: 277–282.

Lisitzin, E., 1965. The mean sea level of the world ocean. *Comment. Phys.–Math. Helsingf.*, 30: 35 pp.

Lisitzin, E., 1967a. Day-to-day variation in sea level along the Finnish coast. *Geophysica*, 9: 259–275.

Lisitzin, E., 1967b. Sea level variations in the Sea of Japan. *Int. Hydrogr. Rev.*, 44: 11–22.

Lisitzin, E., 1969. Les variations saisonnières du niveau de la Mer de Barentz. *Cah. Océanogr. C.O.E.C.*, 21: 673–676.

Lisitzin, E., 1970. The seasonal water balance of the ocean. *Comment. Phys–Math. Helsinf.*, 40: 5 pp.

Lisitzin, E., 1972a. Mean sea level. II. *Oceanogr. Mar. Biol., Ann. Rev.*, 10: 11–25.

Lisitzin, E., 1972b. The complexity of the problem of mean sea level. In: *Studi in Onore di Giuseppina Aliverti*. Istituto Universitario Navale di Napoli, Naples, pp. 157–162.

Lisitzin, E., 1972c. The mean sea level in the Baltic as a function of sea level differences in the transition area. *Conf. Baltic Oceanographers, 8th, Copenhagen*, 7: 4 pp. (duplicated manuscript).

Lisitzin, E., and Pattullo, J., 1961. The principal factors influencing the seasonal oscillations of sea level. *J. Geophys. Res.*, 66: 845–853.

Lundbak, A., 1955. The North Sea storm surge of February 1, 1953. Its origin and development. *Geogr. Tidskr.*, 54: 8–23.

MacMillan, D.H., 1966. *The Tides*. C.R. Books, London, 240 pp.

Malkus, J.S., 1962. Interchange of properties between sea and air: Large scale interaction. In: M.N. Hill (Editor), *The Sea*. Interscience, New York, N.Y. and London, 1: 133, 137.

Masuzawa, J., 1954. On the Kuroshio south off Shiono-Misaki of Japan. *Oceanogr. Mag.*, 6: 25–33.

Matthäus, W., 1972. On the history of recording tide gauges. *Proc. R. Soc. Edinb., B*, 73: 25–34.

Matuzawa, T., 1936. Seismometrische Untersuchungen des Erdbebens vom 2.März 1933. III. Erdbebentätigkeit vor und nach dem Grossbeben. Allgemeines über Nachbeben. *Bull. Earthq. Res. Inst.*, 14: 38–67.

Matuzawa, T., 1937. Directivity of tsunami. *Zisin, Ser. 1*, 9: 23–25 (in Japanese).

Maximov, I.V., 1952. On the "pole tide" in the sea and the atmosphere of the earth. *Dokl. Akad. Nauk SSSR*, 86: 673–676 (in Russian).

Maximov, I.V., 1954. On long period tidal phenomena in the sea and the atmosphere of the earth. *Tr. Inst. Okeanol.*, 8: 18–40 (in Russian).

Maximov, I.V., 1956a. The nutational standing wave in world oceans and its geographical consequences. *Izv. Akad. Nauk SSSR, Ser. Geogr.*, 1: 14–34 (in Russian).

Maximov, I.V., 1956b. The "pole tide" in the oceans of the earth. *Dokl. Akad. Nauk SSSR*, 108: 799–801 (in Russian).

Maximov, I.V., 1958a. Long period variations of the mean sea level in the world oceans. *Nauchn. Dokl. Wyshei Shkoly, (Geol.-Geogr. Nauki)* 3: 11–18 (in Russian).

Maximov, I.V., 1958b. The long period luni-solar tide in the world oceans. *Dokl. Akad. Nauk SSSR*, 118: 888–890 (in Russian).

Maximov, I.V., 1959. The long period luni-solar tides in the seas of the high latitudes of the earth. *Uchonyie Zap. LVIMU.*, 13: 3–38 (in Russian).

Maximov, I.V., 1960. The long period luni-solar tide at the coasts of the Arctic. *Probl. Arct. Antarct.*, 3: 17–20 (in Russian).

Maximov, I.V., 1965a. The experience of the studies of the nine day lunar tide in the Arctic. *Probl. Arct. Antarct.*, 21: 93–96 (in Russian).

Maximov, I.V., 1965b. The solar semi-annual tide in the oceans. *Dokl. Akad. Nauk SSSR.*, 161: 347–350 (in Russian).

Maximov, I.V., 1965c. The solar semi-annual tide in the world oceans. *Probl. Arct. Antarct.*, 21: 11–18 (in Russian).

Maximov, I.V., 1966. The long period luni-solar tides in the oceans. *Okeanologia*, 6: 26–37 (in Russian).

Maximov, I.V., 1967. The water level surface of the oceans and the circulation of the water in the polar zones of the earth. *Probl. Arct. Antarct.*, 27: 169–183 (in Russian).

Maximov, I.V., 1970. *The Geophysical Forces and Water in the Oceans*. Gidrometeorologicheskoie Izdatelstwo, Leningrad, 447 pp. (in Russian).

Maximov, I.V. and Karklin, V.P., 1965. The pole tide in the Baltic Sea. *Dokl. Akad. Nauk SSSR*, 161: 580–582 (in Russian).

Maximov, I.V. and Smirnov, N.P., 1964. The changes in the speed of the earth's rotation and the mean sea level of the oceans. *Okeanologia*, 4: 9–18 (in Russian).

Maximov, I.V. and Smirnov, N.P., 1965. On the origin of the semi-annual rhythm in the activity of the oceanic currents. *Fysika Atmosfery i Okeana*, 1: 1079–1087 (in Russian).

Maximov, I.V., Sarukhanyan, E.I. and Smirnov, N.P., 1970. *Ocean and Kosmos*. Gidrometerologiches-koie Izdatelstwo, Leningrad, 216 pp. (in Russian).

Maximov, I.V., Sarukhanyan, E.I. and Smirnov, N.P., 1972. Long period tidal phenomena in the north part of the Atlantic Ocean. *Cons. Int. Explor, Mer, Rapp. Proc. Verb.*, 162: 285–295.

Maximov, I.V., Vorobjev, V.N. and Smirnov, N.P., 1967. On the study of the nine day lunar tide in the Arctic Ocean. *Okeanologia*, 7: 307–313 (in Russian).

Merz, A. (Bearbeitet von Lotte Möller), 1928. Hydrographische Untersuchungen in Bosporus and Dardanellen. *Veröff. Inst. Meeresk. Berlin, N.R. Geogr.-Naturw. Reihe*, 18: 3–284.

Miller, A.R., 1957. The effect of steady winds on sea level at Atlantic City. *Woods Hole Oceanogr. Inst. Contrib.*, 829: 24–31.

Miyazaki, M., 1965. A numerical computation of the storm surge of the Hurricane Carla 1961 in the Gulf of Mexico. *Oceanogr. Mag.*, 17: 109–140.

Model, F., 1950. Gegenwärtige Küstenhebung im Ostseeraum. *Mitt. Geogr. Ges.*, 49: 64–115.

Montag, H., 1967. Bestimmung rezenter Niveauverschiebungen aus langjährigen Wasserstandsbeobach-tungen der südlichen Ostseeküste. *Arb. Geod. Inst. Potsdam*, 15: 139 pp.

Montgomery, R.B.,1941a. Sea level difference between Key West and Miami, Florida. *J. Mar. Res.*, 4: 32–37.

Montgomery, R.B., 1941b. Transport of the Florida current off Havana. *J. Mar. Res.*, 4: 198–220.

Montgomery, R.B., 1958. Water characteristics of Atlantic Ocean and of world ocean. *Deep-Sea Res.*, 5: 134–148.

Montgomery, R.B., 1969. Comments on oceanic leveling. *Deep-Sea Res.*, 16: 147–152.

Morcos, S.A., 1960. The tidal currents in the southern part of the Suez Canal. In: *Symp. Tidal Estuaries, Int. Assoc. Sci. Hydrolog. – Int. Assoc. Phys. Oceanogr., Publ. 51, I.A.S.H. Comm. Surface Waters*, pp. 307–316.

Morcos, S.A., 1970. Physical and chemical oceanography of the Red Sea. *Oceanogr. Mar. Biol., Ann. Rev.*, 8: 73–202.

Morcos, S.A. and Gerges, M.A., 1968. In: *Arabic Congress for Aquatic Resources and Oceanography*, Ministry Sci. Res., Cairo, 17 pp. (in Arabic).

Morskoi Atlas (The Marine Atlas), II, 1953. Izd.Morskogo Generalnogo Shtaba, Moscow, 24 pp, 76 plates.

Munk, W.H., 1947. A critical wind speed for air–sea boundary processes. *J. Mar. Res.*, 6: 203–218.

Munk, W.H., 1958. The seasonal budget of water. *Geophysics and IGY. Geophys. Monogr., Am. Geophys. Union*, 2: 175–176.

Munk, W.H. and Revelle, R.A., 1952. Sea level and the rotation of the earth. *Am. J. Sci.* 250: 829–833.

Munk, W.H. and Zetler, B.D., 1967. Deep-sea tides: a program. *Science*, 158: 884–886.

Neumann, G., 1941. Eigenschwingungen der Ostsee. *Arch. Dtsch. Seewarte*, 61: 57 pp.

Neumann, G., 1942. Die absolute Topographie des physikalischen Meeresniveaus und die Oberflächen-strömungen des Schwarzen Meeres. *Ann. Hydrogr. Marit. Meteorol.*, 70: 265–282.

Neumann, G., 1948. Über den Tangentialdruck des Windes und der Rauhigkeit der Meeresoberfläche. *Z. Meteorol.*, 2: 193–203.

Newton, I., 1687. *Philosophiae Naturalis Principia Mathematica*. Lib. I, prop. 66, cor. 19; Lib. 3, prop. 24, 26, 37. London.

Nicolini, T., 1950. Indizi di correspondenza tra fenomeni geofisici et moto polare. *Atti Accad. Naz. Lincei*, 8: 348–355, 595–605.

Nomitsu, T., 1934. Coast effect upon the ocean current and sea level, II. Changing state. *Mem. Coll. Sci. Kyoto Univ. A*, 17: 249–280.

Nomitsu, T., 1935. A theory of tsunamis and seiches produced by wind and barometric gradient. *Mem. Coll. Sci. Kyoto Univ., A*, 18: 201–214.

Nomitsu, T. and Okamoto, M., 1927. The causes of the annual variation of the mean sea level along the Japanese coast. *Mem. Coll. Sci. Kyoto Univ., A*, 10: 125–161.

Nowroozi, A.A., Ewing, M., Nafe, J.E. and Fleigel, M., 1968. Deep ocean current and its correlation with the ocean tide off the coast of northern California. *J. Geophys. Res.*, 73: 1921–1932.

Palmen, E., 1936. Über die von einem stationären Wind verursachte Wasserstauung. *V. Hydrol. konf. Balt. Staaten, Ber.* 15B: 17 pp.

Palmén, E., 1967. Evaluation of atmospheric moisture transport for hydrological purposes. *Rep. WMO/IHD Projects*, 1: 63 pp.

Palmén, E. and Laurila, E., 1938. Über die Einwirkung eines Sturmes auf den hydrographischen Zustand im nördlichen Ostseegebiet. *Comment. Phys-Math. Helsingf.*, 10: 53 pp.

Palmén, E. and Söderman, D., 1966. Computation of the evaporation from the Baltic Sea from the flux of water vapor in the atmosphere. *Geophysica*, 1: 261–279.

Pattullo, J., Munk, W., Revelle, R. and Strong, E., 1955. The seasonal oscillations in sea level. *J. Mar. Res.*, 14: 88–156.

Petterssen, S., Bradbury, D.L. and Petersen, K., 1962. The Norwegian cyclone model in relation to heat and cold sources. *Geofys. Publ.*, 24: 243–280.

Pierce, C., 1960. Is the sea level falling or the land rising in S.E. Alaska? *20th Ann. Meet., Am. Congr. Surveying and Mapping*. U.S. Deparment of Commerce, Washington, D.C., 12 pp.

Pobedonoszev, S.V., 1971. The application of the computed mean yearly sea level for the determination of the present movements of the earth's crust. *Geod. Kartogr.*, 3: 18–28 (in Russian).

Pobedonoszev, S.V., 1972. Recent vertical movements of the coastlines of the seas washing the European Territory of the USSR. *Okeanologia*, 12: 741–745 (in Russian).

Pobedonoszev, S.V. and Rosanov, L.L., 1971. Recent vertical movements of the White and the Barents Sea shores (according to mareograph records). *Geomorfologia*, 3: 57–62 (in Russian).

Pollak, M.J., 1958. Frequency distribution of potential temperatures and salinities in the Indian Ocean. *Deep-Sea Res.*, 5: 128–133.

Polli, S., 1942. L'oscillazione annua dell'Oceano Atlantico. *Arch. Oceanogr. Limnogr.*, 2: 199–213.

Polli, S., 1961. Sui periodi delle oscillazioni liberi dell'Adriatico. *Ist. Sperimentale Talassografico, Trieste, Pubbl.*, 380: 183–192.

Polli, S., 1962a. Il problema della sommersione di Venezia. *Atti XII Convegno Assoc. Geofis. Ital.*, 10 pp.

Polli, S., 1962b. Il progressivo aumento del livello marino lungo le coste del Mediterraneo. *Ist. Sperimentale Talassografico, Trieste, Pubbl.*, 390: 649–654.

Proudman, J., 1914. Free and forced longitudinal tidal motion in a lake. *Proc. Lond. Math. Soc.*, 14: 240–250.

Proudman, J., 1928. The determination of earth-tides by means of water-tides in narrow seas. *Bull. Assoc. Océanogr. Phys., UGGI*, 11: 1–6.

Proudman, J., 1929. The effect on the sea of changes in atmospheric pressure and the forced tides in an ocean bounded by a complete meridian on a non-rotating earth. *Mon. Not. R. Astron. Soc., Geophys. Suppl.*, 2: 197–213.

Proudman, J., 1954a. Note on the dynamical theory of storm surges. *Arch. Met. Geophys. Bioklim., A*, 7: 344–351.

Proudman, J., 1954b. Note on the dynamics of storm surges. *Mon. Not. R. Astron. Soc., Geophys. Suppl.*, 7: 44–48.

Proudman, J., 1955a. The effect of friction on the progressive wave of tide and surge in an estuary. *Proc. R. Soc., A*, 233: 407–418.

Proudman, J., 1955b. The propagation of tide and surge in an estuary. *Proc. R. Soc., A*, 231: 8–24.

Proudman, J., 1957. Oscillations of the tide and surge in an estuary of finite length. *J. Fluid Mech.*, 2: 371–382.

Proudman, J., 1958. On the series that represent tides and surges in an estuary. *J. Fluid Mech.*, 3: 411–417.

Proudman, J., 1960. The condition that a long period tide shall follow the equilibrium law. *Bull. Géod.*, 55: 101–102.

Proudman, J., 1963. The storm surge warning service of the east coast and the question of its extension to other coasts. *Dock Harb. Auth.*, 43: 361–362.

Proudman, J. and Doodson, A.T., 1924a. The principal constituents of the tides of the North Sea. *Philos. Trans. R. Soc., A*, 224: 185–219.

Proudman, J. and Doodson, A.T., 1924b. Time-relation in meteorological effects on sea level. *Proc. Lond. Math. Soc., Ser. 2*, 24: 140–149.

Pupkov, V.N., 1964. Formation, distribution and variability of snow cover in the Asiatic territory of the USSR. *Meteorol. Gidrol.*, 8: 34–40 (in Russian).

Reid, J.L., 1961a. On the geostrophic flow of the surface of the Pacific Ocean with respect to the 1,000-decibar surface. *Tellus*, 13: 489–502.

Reid, J.L., 1961b. On the temperature, salinity and density differences between the Atlantic and Pacific Oceans in the upper kilometre. *Deep-Sea Res.*, 7: 265–275.

Robinson, A.H.W., Barnes, F.A., King, A.M., Edwards, K.C., Grove, A.T., MacGregor, D.R., Boerman, W.E. and Ougton, M., 1953. The storm floods of 1st Februari 1953. *Geography*, 134–189.

Roden, G.I., 1963. Sea level variations at Panama. *J. Geophys. Res.*, 68: 5701–5710.

Romanovsky, V., 1954. Les courants marins de surface dans le bassin occidental de la Méditerranée. *Assoc. Océanogr. Phys., Proc. Verb.*, 6: 256–257.

Rossiter, J.R., 1954a. Report on the investigation of the secular variation of sea level on the coasts of the British Isles, the Canaries and Azores, Egypt, the Gold Coast, Australia, and on the shores of the Indian Ocean, also for Poland. *Assoc. Océanogr. Phys., Publ. Sci.*, 13: 16–21.

Rossiter, J.R., 1954b. The North Sea storm surge of 31 January and 1 February 1953. *Philos. Trans. R. Soc., A*, 246: 371–400.

Rossiter, J.R., 1959. A method for extracting storm surges from tidal records. *Dtsch. Hydrogr. Z.*, 12: 117–127.

Rossiter, J.R., 1960. Report on the reduction of sea level observations (1940–1958) for REUN. *Trav. Assoc. Int. Géod.*, 21: 159–184.

Rossiter, J.R., 1961. Interaction between tide and surge in the Thames. *Geophys. J.*, 6: 29–53.

Rossiter, J.R., 1963. Tides. *Oceanogr. Mar. Biol., Ann. Rev.*, 1: 11–25.

Rossiter, J.R., 1967. An analysis of annual sea level variations in European waters. *Geophys. J.*, 12: 259–299.

Rossiter, J.R., 1968. Mean sea level research. *Conf. Commonwealth Survey Officers, London, 1967*, pp. 133–138.

Rossiter, J.R., 1972a. Sea-level observations and their secular variation. *Philos. Trans. R. Soc.*, 272: 131–139.

Rossiter, J.R., 1972b. The history of tidal predictions in the United Kingdom before the twentieth century. *Proc. R. Soc. Edinb., B*, 73: 13–23.

Rossiter, J.R. and Lennon, G.W., 1965. Computations of tidal conditions in the Thames estuary by initial value method. *Proc. Inst. Civ. Engineers*, 31: 25–56.

Rouch, J., 1944. La variation du niveau de la mer en fonction de la pression atmosphérique d'après les observations du "Pourquoi Pas? " dans l'Antarctique. *Bull. Inst. Océanogr. Monaco*, 870: 1–7.

Rouch, J., 1948. *Traité d'Océanographie Physique. Les Mouvements de la Mer.* Payot, Paris, 413 pp.

Rybak, B.H., 1971. On the computation of the sea level and the currents in the White Sea during storm surges. *Gl. Upr. Gidrometeorol. Slushby, Tr.*, 83: 41–48 (in Russian).

Rybak, B.H., 1972. The hydrodynamical computation of the sea level in the White Sea. *Gl. Upr. Gidrometeorol. Slushby. Tr. Vses. Konf. Molodych Uch Gidromet Slushby SSSR.* (The Results of the All-Union Konference the the Young Scientists of the Hydromet-service of the USSR), pp. 29–34 (in Russian).

Sager, G., 1959. *Ebbe und Flut.* Hermann Haack, Gotha, 251 pp.

Schureman, P., 1924. *A Manual of the Harmonic Analysis and Prediction of Tides. U.S. Coast Geeod. Surv., Spec. Publ.*, 98, 416 pp.

Schweydar, W., 1916. Theorie der Deformation der Erde durch Flutkräfte. *Veröff. Preuss. Geod. Inst., Neue Folge*, 66: 55 pp.

Shimizu, T., 1963. The variation of the sea level and the barometric pressure with Chandler's period. *Geophys. Pap. dedicated to Prof. Kenso Sassa*, pp. 499–515.

Shumsky, P.A., Krenke, A.N. and Zotikov, I.A. 1964. Ice and its changes. In: *Research in Geophysics*. M.I.T. Press, Cambridge, Mass., 2: 425–460.

Sieger, R., 1893. Seenschwankungen und Strandverschiebungen in Skandinavien. *Z. Ges. Erdkd., Berlin*, 28: 393–488.

Simojoki, H., 1948. On the evaporation from the Northern Baltic. *Geophysica*, 3: 123–126.

Simojoki, H., 1949. Niederschlag und Verdunstung auf dem Baltischen Meer. *Fennia*, 71: 25 pp.

Snodgrass, F., 1968. Deep-sea instrument capsule. *Science*, 162: 78–87.

Soloviev, S.L., 1972. The methods for tsunami forecast. *Vestn. Akad. Nauk SSSR*, 5: 72–81 (in Russian).

Soskin, I.M. and Rosova, L.V., 1957. The water exchange between the Baltic and the North Sea. *Tr. Gos. Okeanogr. Inst.*, 41: 9–30 (in Russian).

Starr, V.P. and Peixoto, J.P., 1958. On the global balance of water vapour and the hydrology of the deserts. *Tellus*, 10: 189–194.

Starr, V.P., Peixoto, J.P. and Crisi, A.R., 1965. Hemispheric water balance for the IGY. *Tellus*, 17: 463–472.

Steers, J.A., 1953. The East coast floods, Jan. 31 – Febr. 1, 1953. *Geogr. J.*, 119: 280–289.

Stenij, S.E. and Hela, I., 1947. Suomen merenrannikoiden vedenkorkeuksien lukuisuudet. (English summary: Frequency of the water heights on the Finnish coast.) *Merentutkimuslaitoksen Julkaisu-Havsforskningsinstitutets Skrift*, 138: 21 pp.

Stommel, H., 1964. Summary charts of the mean dynamic topography and current field at the surface of the ocean, and related functions of the mean wind stress. In: *Studies on Oceanography, dedicated to Prof. Hidaka*, pp. 53–58.

Stommel, H., 1965. *The Gulf Stream, a Physical and Dynamical Description*. Univ. Calif. Press, Berkeley, Calif., 2nd ed., 248 pp.

Stoneley, R., 1967. Tsunami. *Int. Dictionary Geophysics*, 2: 1598–1603.

Stovas, V.M., 1951. *On the Question of the Critical Parallels of the Earth Ellipsoid*. Thesis, Univ. of Leningrad, Leningrad (in Russian).

Sturges, W., 1967. Slope of sea levels along the Pacific coast of the United States. *J. Geophys Res.*, 72: 3627–3637.

Sturges, W., 1968. Sea surface topography near the Gulf Stream. *Deep-Sea Res.*, 15: 149–156.

Svansson, A., 1959. Some computations of water heights and currents in the Baltic. *Tellus*, 2: 231–238.

Svansson, A., 1972. On the water exchange of the Baltic. *Conf. Baltic Oceanographers, 8th, Copenhagen, Pap.*, 26: 10 pp. (duplicated manuscript).

Sverdrup, H.U., 1933. Vereinfachtes Verfahren zur Berechnung der Druck- und Massenverteilung im Meere. *Geophys. Publ.*, 10: 1–9.

Sverdrup, H.U., 1957. Oceanography. In: S. Flügge, Editor, *Handbuch der Physik – Encyclopedia of Physics, 48, Geophysics*, 2: 608–670.

Sverdrup, H.U., Johnson, M.W. and Fleming, R.H., 1942. *The Oceans, their Physics, Chemistry and General Biology*. Prentice Hall, New York, N.Y., pp. 400–575, 605–761.

Thomsen, H., 1963. Danish observations in Great Belt and the Baltic. *Cons. Int. Explor. Mer, Ann. Biol.*, 20: 62.

Thomsen, H. and Hansen, B., 1970. Middelvandstand og dens aendring ved de Danske kyster. *Danske Meteorol. Inst., Medd.*, 23: 24 pp.

Thorarinsson, S., 1940. Present glacier shrinkage and eustatic changes in sea level. *Geogr. Ann.*, 12: 131–159.

Tsumura, K., 1963. Investigation of the mean sea level and its variation along the coast of Japan (Part 1) – Regional distribution of sea level variation. *J. Geod. Soc. Japan*, 9: 49–90.

Uda, M., 1971. Oceanic study in Japan – its progress in the last decade. *J. Oceanogr. Soc. Japan*, 27: 233–235.

United European Leveling Net (UELN), 1959. Int. Assoc. Geod., Copenhagen.

Uusitalo, S., 1960. The numerical calculations of wind effect on sea level elevations. *Tellus*, 12: 427–435.

Uusitalo, S., 1972. Numerical investigation of the influence of wind on water levels and currents in the Gulf of Bothnia. A preliminary experiment. *Merentutkimuslaitoksen Julkaisu-Havsforskningsinstitutets Skrift*, 235: 25–68.

Vadati, K., 1967. Tsunami. *Int. Dictionary Geophysics*, 2: 1598–1603.

Van Dantzig, D. and Lauwerier, H.A., 1960a. General considerations concerning the hydrodynamical problem of the motion of North Sea, The North Sea Problem I. *Proc. K. Ned. Akad. Wet., A*, 63: 170–180.

Van Dantzig, D. and Lauwerier, H.A., 1960b. Free oscillation of a rotating rectangular sea, The North Sea Problem IV. *Proc. K. Ned. Akad. Wet., A*, 63: 339–354.

Van Dorn, W.G., 1961. Some characteristics of the surface gravity waves in the sea produced by nuclear explosions. *J. Geophys. Res.*, 66: 3845–3862.

Van Hylckama, T.E.A., 1956. The water balance of the earth. In: *Publications in Climatology*. Drexel Inst. Techn. Lab. Climatol., Centerton, N.J., 57–117.

Veronis, G. and Stommel, H., 1956. The action of variable wind stresses on a stratified ocean. *J. Mar. Res.*, 15: 43–75.

Voipio, A., 1964. Salinity variations as the indicator of the rate of water transport along the east coast of the Bothnian Sea. *Geophysica*, 9: 49–63.

Voit, S.S., 1956. *What are the Tides?* Izdatelstwo Akademii Nauk SSSR, Moscow, 102 pp. (in Russian).

Von Sterneck, R., 1914. Über Seiches an der Küste der Adria. *Sitz. Ber. Akad. Wiss. Wien*, 123: 2199–2232.

Von Sterneck, R., 1920. Die Gezeiten der Ozeane. *Sitz. Ber. Akad. Wiss. Wien*, 129: 131–150.

Von Sterneck, R., 1927. Selbstständige Gezeiten und Mitschwingungen im Roten Meer. *Ann. Hydrogr. Marit. Meteorol.*, 55: 129–144.

Waldichuk, K.M., 1964. Daily and seasonal sea level oscillations on the Pacific coast of Canada. In: *Studies on Oceanography, dedicated to Prof. Hidaka*, pp. 181–201.

Weenink, M.P.H., 1958. A theory and method of calculation of wind effects on sea levels in a partly-enclosed sea, with special application to the southern coast of the North Sea. *Medel. Verh. K. Ned. Meteorol. Inst.*, 73: 111 pp.

Weenink, M.P.H. and Groen, P., 1958. A semi-theoretical, semi-empirical approach to the problem of finding wind effects on water levels in a shallow, partly-enclosed sea. I and II. *Proc. K. Ned. Akad. Wet., B*, 61: 198–213.

Welander, P., 1957. Wind action on a shallow sea: Some generalization of Ekman's theory. *Tellus*, 9: 45–52.

Welander, P., 1961. Numerical prediction of storm surges. In: H.E. Landsberg and J. van Mieghem (Editors), *Advance in Geophysics*. Academic Press, New York, N.Y., and London, pp. 315–379.

Wemelsfelder, P.J., 1953. The disaster in The Netherlands caused by the storm flood of 1 February 1953. *Proc. Conf. Coast. Eng., 4th, Berkeley, Calif.*, pp. 258–271.

Wexler, H., 1961. Ice budget for Antarctica and changes of sea level. *J. Glaciol.*, 3: 867–872.

Witting, R., 1911. Tidvattnet i Ostersjön och Finska viken. *Fennia*, 29: 84 pp.

Witting, R., 1918. Hafsytan, geoidytan och landhöjningen utmed Baltiska hafvet och vid Nordsjön. *Fennia*, 39: 347 pp.

Witting, R., 1922. Le soulèvement récent de la Fennoscandie. Quelques mots à propos de l'article de M. Rune dans ces annales. *Geogr. Ann.*, 4: 458–487.

Witting, R., 1943. Landhöjningen utmed Baltiska havet under åren 1898–1927. *Fennia*, 68: 40 pp.

Wunsch, C., Hansen, D.V. and Zetler, B.D., 1969. Fluctuations of the Florida Current inferred from sea level records. *Deep-Sea Res.*, 16 (Suppl.): 447–470.

Wüst, G., 1920. Die Verdunstung auf dem Meere. *Veröff. Inst. Meeresk., Berlin*, 6: 1–95.

Wüst, G., 1954. Gezetzmässige Wechselbeziehungen zwischen Ozean und Atmosphere, in der zonalen Verteilung von Oberflächensalzgehalt, Verdunstung und Niederschlag. *Arch. Meteorol. Geophys. Bioklim. (Defant Festschr.)., A*, 7: 305–328.

Wyrtki, K., 1954. Der grosse Salzeinbruch in die Ostsee in November und December 1951. *Kieler Meeresforsch.*, 10: 19–25.

Yamaguti, S., 1962. On the changes in the heights of mean sea level at San Francisco. *J. Oceanogr. Soc. Japan*, 20: 159–167.

Yamaguti, S., 1965. On the changes in the heights of mean sea levels before and after the great Niigata earthquake on June 16, 1964. *J. Geod. Soc. Japan*, 10: 187–191; *Bull. Earthq. Res. Inst.*, 45: 167–172.

Zetler, B.D., 1971. Radiational ocean tides along the coasts of the United States. *J. Phys. Oceanogr.*, 1: 34–38.

Zetler, B.D. and Cummings, R.A., 1967. A harmonic method for prediction shallow-water tides. *J. Mar. Res.*, 25: 103–114.

Zetler, B.D. and Maul, G.A., 1971. Precision requirements for a spacecraft tide program. *J. Geophys. Res.* 76: 6601–6605.

Zubov, N.N., 1959. The influence of baric relief on sea level and currents. In: *International Oceanographic Congress, 1959. Am. Assoc. Adv. Sci., Preprints*: 792–796.

AUTHOR INDEX

SUBJECT INDEX